Contributions from Science Education Research

Volume 5

Series Editor
Robin Millar, University of York, UK

Editorial Board
Costas P. Constantinou, University of Cyprus, Nicosia, Greece
Justin Dillon, Exeter University, UK
Reinders Duit, University of Kiel, Germany
Doris Jorde, University of Oslo, Norway
Dimitris Psillos, Aristotle University of Thessaloniki, Greece
Andrée Tiberghien, University of Lyon II, France

More information about this series at http://www.springer.com/series/11180

Olia E. Tsivitanidou • Peter Gray • Eliza Rybska
Loucas Louca • Costas P. Constantinou
Editors

Professional Development for Inquiry-Based Science Teaching and Learning

 Springer

Editors
Olia E. Tsivitanidou
Department of Educational Sciences
University of Cyprus
Nicosia, Cyprus

Peter Gray
Norwegian University of Science
and Technology
Trondheim, Norway

Eliza Rybska
Department of Nature Education
and Conservation
Adam Mickiewicz University
Poznań, Poland

Loucas Louca
Department of Education Sciences
European University Cyprus
Nicosia, Cyprus

Costas P. Constantinou
Department of Educational Sciences
University of Cyprus
Nicosia, Cyprus

ISSN 2213-3623   ISSN 2213-3631   (electronic)
Contributions from Science Education Research
ISBN 978-3-030-08245-1   ISBN 978-3-319-91406-0   (eBook)
https://doi.org/10.1007/978-3-319-91406-0

Printed on acid-free paper

This Springer imprint is published by the registered company Springer International Publishing AG part of Springer Nature.
The registered company address is: Gewerbestrasse 11, 6330 Cham, Switzerland

Contents

What Is Inquiry-Based Science Teaching and Learning?   1
Costas P. Constantinou, Olia E. Tsivitanidou, and Eliza Rybska

Part I Promoting Student Inquiry in the Science Classroom

**Science Inquiry as Part of Technological Design:
A Case of School-Based Development in Norway** 27
Berit Bungum

**Promoting IBSE Using Living Organisms: Studying Snails
in the Secondary Science Classroom** 43
Eliza Rybska

Drama As a Learning Medium in Science Education 65
Ran Peleg, Anna-Lena Østern, Alex Strømme, and Ayelet Baram Tsabari

**Part II Familiarizing Teachers with Motivational Approaches
and Scientific Literacy Goals for Inquiry Based Learning**

Using Motivational Theory to Enrich IBSE Teaching Practices 87
Hanne Møller Andersen and Lars Brian Krogh

**Taking Advantage of the Synergy Between Scientific Literacy
Goals, Inquiry-Based Methods and Self-Efficacy
to Change Science Teaching** 105
Robert Evans and Jens Dolin

Inquiry-Based Approaches in Primary Science Teacher Education 121
Sami Lehesvuori, Ilkka Ratinen, Josephine Moate, and Jouni Viiri

**Part III   Fostering Teachers' Competences in Cross-Domain
Scientific Inquiry**

Promoting Pre-service Teachers' Ideas About Nature of Science
Through Science-Related Media Reports.   137
Gultekin Cakmakci and Yalcin Yalaki

The Development of Collaborative Problem-Solving Abilities
of Pre-service Science Teachers by Stepwise
Problem-Solving Strategies   163
Palmira Pečiuliauskienė and Dalius Dapkus

Teachers as Educational Innovators in Inquiry-Based Science
Teaching and Learning.   185
Anni Loukomies, Kalle Juuti, and Jari Lavonen

**Part IV   Capitalizing on Teacher Reflections to Enhance
Inquiry-Based Science Teaching**

The Biology Olympiad as a Resource and Inspiration
for Inquiry-Based Science Teaching   205
Jan Petr, Miroslav Papáček, and Iva Stuchlíková

A Teacher Professional Development Programme
on Dialogic Inquiry.   223
Margareta Enghag, Susanne Engström, and Birgitta Norberg Brorsson

Designing Teacher Education and Professional Development
Activities for Science Learning   245
Andrée Tiberghien, Zeynab Badreddine, and David Cross

Concluding Remarks: Theoretical Underpinnings in Implementing
Inquiry-Based Science Teaching/Learning.   261
Loucas Louca, Thea Skoulia, Olia E. Tsivitanidou,
and Costas P. Constantinou

Index.   281

What Is Inquiry-Based Science Teaching and Learning?

Costas P. Constantinou, Olia E. Tsivitanidou, and Eliza Rybska

1 Introduction

Inquiry is the intentional process of diagnosing situations, formulating problems, critiquing experiments and distinguishing alternatives, planning investigations, researching conjectures, searching for information, constructing models, debating with peers using evidence and representations and forming coherent arguments (Linn, Davis, & Bell 2004). Inquiry has been a strongly advocated approach to teaching and learning generally and particularly in science for many years. It refers to a learning process in which students are actively engaged (Anderson, 2002). Inquiry has implications for designing learning environments, for planning teaching and for assessing students' learning achievements.

Inquiry in science education has a long and complex history: the term "inquiry" is grounded in the ideas of key educators such as Dewey (1996) and Bruner (1960). John Dewey has rooted inquiry in experience and has described the pattern of inquiry, which is located in human culture, language and everyday experience. According to Dewey, learning experiences should be collaborative and placed in a frame of reconstruction of knowledge. Dewey also highlighted the role of reflection. While describing practical forms of inquiry, he included three situations: pre-reflection, reflection and post-reflection (Dewey, 1933 as cited in Garrison, Anderson, & Archer, 1999). The process of going beyond the information provided, according to Bruner (1960) represents a specific educational procedure that resembles the scientific inquiry

C. P. Constantinou (✉) · O. E. Tsivitanidou
Department of Educational Sciences, University of Cyprus, Nicosia, Cyprus
e-mail: c.p.constantinou@ucy.ac.cy; tsivitanidou.olia@ucy.ac.cy

E. Rybska
Department of Nature Education and Conservation, Adam Mickiewicz University, Poznań, Poland
e-mail: elizaryb@gmail.com

© Springer International Publishing AG, part of Springer Nature 2018
O. E. Tsivitanidou et al. (eds.), *Professional Development for Inquiry-Based Science Teaching and Learning*, Contributions from Science Education Research 5, https://doi.org/10.1007/978-3-319-91406-0_1

process. This process requires retrospective (changing schemas, data management, exploring meaning) and prospective (formulating new hypothesis) ways of thinking which constitute essential features of inquiry learning (Filipiak, 2011).

As with any attempt at defining the term "inquiry", likewise it is hard to trace exactly the first appearance of "inquiry instruction", even though we can argue that inquiry instruction has its origins on the long-standing dialogue about the nature of learning and teaching and particularly the work of Jean Piaget (with his idea of conceptualization, knowledge construction and the role of experience), Lev Vygotsky (with his social meaning of learning, and his reference to the method of scientific cognition, where understanding and reasoning become key elements in this process; see Vygotsky, 1971) and David Ausubel (with his work learning by discovery and his work on meaningful learning considering learners' prior knowledge; see Ausubel, 1961, 2012). The work of these theorists was inter-mingled into the philosophy of learning known first as constructivism (Cakir, 2008) and in a developed form as social constructivism (Mayer, 2004), which were both used to shape instructional materials and overall to reconceptualize science teaching and learning.

The early years of science education in comprehensive educational systems throughout the world, which sought to take in all students, were burdened by expert-/teacher-centred transmissive approaches with a sole emphasis on presenting established knowledge (as a homogeneous, undisputed construct) for the small fraction of students to assimilate, if they were interested in developing disciplinary expertise from an as young age as possible (Rocard et al., 2007). The advent of science education research with the clear identification of misconceptions and alternative reasoning approaches (e.g. Driver, Guesne, & Tiberghien, 1985) as well as the first results of international educational assessments led to the child-centred discovery approaches with emphases on process-based learning and scientific methods. These approaches were not new, but they were not yet grounded to research data as the initiatives for research in this field were just initiated. This was the first failed attempt to infuse authenticity to the science classroom by instilling what was thought to be the essence of laboratory practice in natural science research (Chinn & Malhotra, 2002).

The theoretical debate that followed this failure led to new perspectives on constructivist and social constructivist or constructionist approaches. Although the main problem was not solved, as a range of constructivist approaches sought to infuse educational practice in the 1970s, the focus on investigation practices and, later, inquiry approaches became particularly prominent in science education (Minner, Levy, & Century, 2010).

What needs to be mentioned here is that for somebody unfamiliar with science education topics inquiry may seem to be used as synonym for constructivism. Constructivism is a theory of learning, a theory that explains how construction of knowledge occurs in a person's mind (Klus-Stańska, 2000; Michalak, 2004). IBSE encompasses features of constructivism but is not limited to this. However, inquiry should not be confused with constructivism, and there has been considerable debate as to how far constructivism extends into inquiry, or vice versa. According to Osborne: "Four decades after Schwab's (1962) argument that science should be taught as an 'enquiry into enquiry', and almost a century since John Dewey (1916) advocated that classroom learning be a student-centred process of enquiry, we still find ourselves struggling to achieve such practices in the science classroom" (Osborne, 2006, p. 2). Moreover, inquiry is often confounded or used alternatively with other terms that describe more specific learning and teaching approaches such as *anchored instruction, hands-on, problem-based, project-based, student-centered, inductive* and *dialogic approaches* (Anderson, 2002; Hayes, 2002).

Despite the fact that there is no generally accepted clear definition of inquiry-based education, common core elements do exist and are discussed later in this chapter.

2   What Is Inquiry-Based Science Education (IBSE)?

Historically, two pedagogical approaches in science teaching can be contrasted: (1) deductive and (2) inductive approaches. In deductive or so-called top-down transmission approaches, teachers' role was confined to presenting the scientific concepts and their logical – deductive – implications and to giving examples of applications, whereas learners, as passive receivers of knowledge, were forced to handle abstract notions. The inductive or so-called bottom-up approaches gave space to observation, experimentation and the teacher-guided construction by the learners of their own knowledge (Rocard et al., 2007). According to Rocard et al.'s (2007) report, "The terminology evolved through the years and the concepts refined, and today the Inductive Approach is most often referred to as Inquiry-Based Science Education (IBSE), mostly applied to science of nature and technology" (p. 14).

The last two decades have seen growing calls for inquiry to play an important role in science education (e.g. Blumenfeld et al., 1991; Linn, Pea, & Songer, 1994; National Research Council, 1996; Rocard et al., 2007). This call for inquiry-based learning is based on the recognition that science is essentially a question-driven, open-ended process of constructing coherent conceptual frameworks with predictive capabilities and that students must have personal experience with scientific inquiry and engage in its practices, in order to be enculturated in these fundamental aspects of science (Linn, Songer, & Eylon, 1996; NRC, 1996).

However, one difficulty for efforts to promote inquiry is the lack of specificity of what it can mean, in classroom terms. Other researchers (Anderson, 2002; Minner et al., 2010) have discussed this problem of ambiguity in the term inquiry and described three distinct meanings of the term in the literature (see Fig. 1): (1) *scientific inquiry*, referring to the diverse ways in which scientists practise to generate and validate knowledge; (2) *inquiry learning*, referring to the active learning processes in which students are inevitably engaged; and (3) *inquiry teaching*, which is the main focus of literature around inquiry, for which there is no clear operational definition.

What is worth mentioning is that the educational process by itself consists of two major actors: the teacher and the learner(s). Hence, it involves two processes, namely, teaching and learning, which may rely on different methods, strategies and principles. The educational process has a cognitive as well as a cultural facet,

Fig. 1 Three distinct
perspectives of the term
"inquiry" in the literature
(Anderson, 2002; Minner
et al., 2010)

applied through communications among the different actors. In the sections to follow
in this chapter, we distinguish between inquiry-based science learning (IBSL) and
inquiry-based science teaching (IBST) in order to provide a holistic overview of the
notion of inquiry in science learning.

2.1 Inquiry-Based Science Learning (IBSL)

In any of the three perspectives discussed by Anderson (2002) and Minner et al.
(2010), namely, whether we refer to scientists, students or teachers who do inquiry,
some core components characterize those enactments. From the learners' perspective, those core components are described by the National Research Council (NRC)
as "essential features of classroom inquiry" (NRC, 2000, p. 25 as cited in Minner
et al. 2010), including (1) learners being engaged by meaningful scientifically oriented questions; (2) learners giving priority to evidence, which allows them to
develop and evaluate ideas that address scientific questions; (3) learners formulating
knowledge claims and arguments from evidence in order to settle scientific questions; (4) learners evaluating their explanations in light of alternative explanations,
particularly those reflecting scientific understanding; and (5) learners communicating and justifying their proposed explanations.

According to Arnold, Kremer, and Mayer (2014), while learners engage in
inquiry as a means, they are supposed to also learn scientific content knowledge
through inquiry. Since in such lens inquiry leads to knowledge construction, thus
in this vein, "inquiry" can be also seen as an outcome. Students learn how to do
science and acquire relevant skills or abilities, and they develop an understanding
of scientific inquiry itself (NRC, 1996). With the introduction of the Next
Generation Science Standards (NGSS Lead States, 2013), there has been a shift
from the notion of "inquiry skills" to the notion of "science practices" (Bybee,

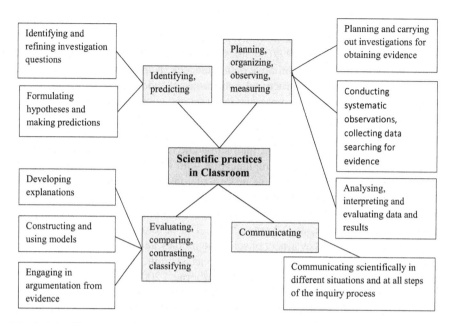

Fig. 2 Scientific practices fostered in classroom inquiry in science education (NRC, 2000)

2011). The term "practices" is meant "to stress that engaging in scientific inquiry requires coordination both of knowledge and skill simultaneously" (NRC, 2012, p. 41). With this respect, the process of scientific inquiry in science education involves the development of an understanding of scientific aspects of the world around through identifying and refining investigation questions; formulating hypotheses and/or making predictions; planning, managing and carrying out investigations with a purpose to obtain evidence (e.g. conducting systematic observations for searching for relevant evidence); analysing and evaluating data; interpreting results; developing explanations; constructing and using models; engaging in argumentation from evidence; and being able to communicate scientifically in different situations and at all steps of the inquiry process (see Fig. 2).

Alongside the acquisition of scientific practices (Bybee, 2011) and an understanding of scientific concepts and phenomena (Schroeder, Scott, Tolson, Huang, & Lee, 2007), classroom inquiry also fosters learners' thinking skills and critical thinking (Haury, 1993), offers experiences with science, promotes the development of an epistemological awareness of how science operates (Chinn & Malhotra, 2002) and develops positive attitudes towards science (Shymansky, Kyle, & Alport, 1983). Moreover, the acquisition of core practices, such as modelling and argumentation, is deemed essential for responsible citizenship and success in the twenty-first century (Beernaert et al., 2015; Pellegrino & Hilton, 2013). Inquiry also provides the opportunity to acquire specific investigation skills, relying on different methods of investigation and different sources of evidence. From an educational perspective, the following forms of inquiry have been proposed by researchers: controlled experimentation (Schauble, Glaser, Duschl, Schulze, & John, 1995), modelling (Jackson, Stratford, Krajcik, & Soloway, 1996; Penner, Giles, Lehrer, & Schauble, 1997; Wilensky & Resnick, 1999), synthesis of primary sources (Linn, Bell, & Hsi, 1998) and exploration of quantitative data (Hancock, Kaput, & Goldsmith, 1992). All these forms constitute structured collections of evidence from systems and involve the use of evidence to represent, interpret and communicate credibility.

In IBSE, modelling and argumentation constitute key practices that need to be fostered at all educational levels (NRC, 2012). Inquiry itself can promote a culture of collaborative group work, a peer interaction and consequently a construction of discursive argumentation and communication with others as the main process of learning. Argumentation refers to the process of constructing and negotiating arguments (Osborne, Erduran, & Simon, 2004), either individually or cooperatively, which can be expressed either verbally in discussions or any oral statements or in writing (Driver, Newton, & Osborne, 2000). The development of argumentation skills is recognized as a key aspect of scientific literacy and is widely recognized as an important practice for citizenship and also as a significant learning objective of science teaching (Erduran, Simon, & Osborne, 2004; Jiménez-Aleixandre, Rodriguez, & Duschl, 2000; NRC, 2012). The other core practice in science education, which is also important in inquiry, is modelling (Beernaert et al., 2015; Pellegrino & Hilton, 2013).

Modelling is conceptualized as a process of constructing and deploying scientific models (Hodson, 1993). The development of modelling practices is thought to also facilitate student learning of science concepts, methodological processes and the development of an awareness of how science operates (Hodson, 1993; Saari & Viiri, 2003; Schwarz & White, 2005). Moreover, learners communicate adequate evidence in supporting scientific claims and constructing scientific explanations while modelling a phenomenon. With the presence of appropriate scaffolding, learners can develop evidence-based reasoning and construct scientific explanations (Kyza, Constantinou, & Spanoudis, 2011).

Overall inquiry learning processes are thought to be powerful in developing scientific literacy, since it involves such practices as experimenting, argumentation, modelling, reasoning, etc. All these aspects are deemed important for understanding environmental, medical, economic and other issues that confront modern societies, which rely heavily on technological and scientific advances of increasing complexity (Rocard et al., 2007).

2.2   Inquiry-Based Science Teaching (IBST)

Inquiry is used in a variety of ways with respect to teaching. As inquiry learning is recognized by academics, teachers and practitioners as vital in science learning and children's development overall, it is expected that it will be also prominent in science teaching, without implying that in this context one unique teaching approach may be pursued in science education. In Anderson's scheme, inquiry pedagogies emphasize, among others, the teacher's role: a shift from "dispenser of knowledge" to facilitator or coach for supporting students' learning (Anderson, 2002). Therefore, the role of the teacher switches from being the authority to becoming a guide who challenges students to think beyond their current processes by offering guided questions (Windschitl, 2002) and/or preparing wisely planned scaffolds.

Teachers' capabilities on orchestrating and facilitating inquiry-oriented learning processes are essential. These capabilities cover issues such as efficacy (e.g. efficacy for instructional strategies, for classroom management and/or for student engagement), teacher motivation and enthusiasm for teaching (Tschannen-Moran & Hoy, 2001). It seems that one of the central strategies for teaching science (in agreement with the idea of teaching for meaningful learning proposed by Mayer (2002) is involving students in inquiry activities with questions that are meaningful to them (e.g. generated from their own experiences) and with the explicit aim to develop coherent knowledge and rigorous understanding of phenomena, as well as understanding of how scientists study the natural world and what ideas they have developed in the process. For achieving that, the teacher needs to prepare an ingenious and planned scaffolding, for assisting the students through modelling and coaching in particular by the use of questioning strategies (Barrow, 2006; Prince & Felder, 2007). The teacher also facilitates appropriate discussion and helps students to focus on experimental data and facts, for example, by highlighting the purpose of the experimentation (Baker & Leyva, 2003), by using formative assessment methods or simply by asking meaningful questions (such as, e.g. in QtA discussion – see Beck, 1997).

Considering the fact that IBST has brought fundamental changes in several aspects of pedagogy (Harlen, 2013), as well as the main dimension of IBST and science inquiry, Grangeat (2016) presents and evaluates a six-dimensional model describing the different modalities of inquiry-based teaching. The six dimensions upon which the model is built represent the crucial characteristics of IBST: (1) the origin of questioning, (2) the nature of the problem, (3) students' responsibility in conducting the inquiry, (4) the management of student diversity, (5) the role of argumentation and (6) the explanation of the teacher's goals. This six-dimensional model of IBST might be of value to researchers and teacher educators who are confronted with the complexity of inquiry-based science teaching. Evaluation of the model with qualitative data from secondary science teachers' teaching practices has stressed, among others, the role of formative assessment within inquiry-based teaching as a way to support students in understanding teachers' goals and monitoring their own progress towards those learning goals (Grangeat, 2016).

Although the concept of IBSE has been widely described and in some cases over that last years was also adopted in the European periphery, its defining features are not brand new in many European educational systems. Even though concepts, such as problem-based learning, project-based instruction, inductive thinking, critical thinking, experiential learning and scientific method of learning, are already familiar to many teachers, the concept of IBSE seems to them as rather distant. Teachers confront difficulties in understanding what is expected from them when asked to

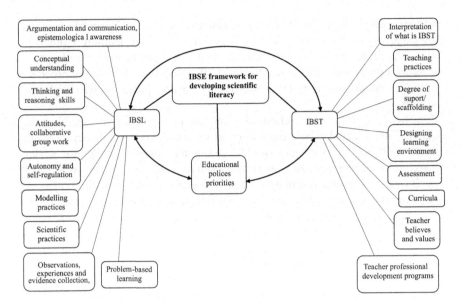

Fig. 3 The main features of the IBSE (inquiry-based science education) framework; *IBSL* inquiry-based science learning, *IBST* inquiry-based science teaching

teach science by inquiry, and to the same extent, this confusion is understandable as there are many definitions on inquiry-based learning (Corbett, 2014). It is therefore reasonable to bridge the gap between newly introduced concepts and current teachers' experience through practices with which the teachers are currently familiar, albeit as an extracurricular activity. The interpretation of what constitutes IBSL and IBST for the community of science educators is crucial for the practices to be endorsed and the principles to be satisfied while designing inquiry-based learning environments.

Similarly, the discussion about IBSE features is linked to the assessment methods and the teacher professional learning and professional developments proposed and implemented. Likewise, educational policy priorities and recommendations from expert groups have an impact on the transformation of the pedagogies promoted in science education. All these comprise essential features of the IBSE framework that we further discuss in this chapter (see Fig. 3).

2.3   The Notion of Inquiry-Based Science Teaching and Learning (IBST/L) in This Book

Inquiry-based science education has a long history, and there are many approaches to teaching and learning science as inquiry (Barrow, 2006; Prince & Felder, 2007).

Our use of the term "Inquiry-based Science Teaching and Learning" (IBST/L) in this book reflects one segment in a historical cycle, a segment where inquiry in science and mathematics education has been seen as an answer to pressing societal problems, including disengagement from science subjects and the need for citizens to be able to debate pressing socio-scientific issues from an informed position. From this perspective, IBST/L brings vital features of authentic scientific inquiry into the science classroom.

In the S-TEAM project, from which the list of chapters included in this book has resulted, a large and diverse group of science educators, researchers and teacher educators came together bringing a variety of interpretations of inquiry to the table. These ranged from procedural interpretations, including ideas about teaching sequences, through ideas about key practices and competences, to what we might call open interpretations, where teacher and pupil empowerment was seen as the central goal.

Our way of dealing with the existing diversity in IBST/L was to embrace it, rather than create a single definition of inquiry. A simple principle, however, runs through all the activities undertaken or promoted by the project. The principle is: "Don't give the answer in advance". In other words, inquiry involves a degree of autonomy or so-called alternative interpretation, connected at the same time with responsibility for learning, given to learners or teachers in progressing from a question, or a problem situation, or one level of knowledge about a topic, towards an answer, a solution or a higher level of knowledge.

Within this principle, there are a large number of variables, or dimensions, and some of our colleagues have constructed five- or six-dimensional models of inquiry as perceived by teachers. Within these models, inquiry is a multifaceted activity; it involves a process of exploring phenomena, asking questions, investigating, representing and reasoning to construct explanations and new understandings of the world. Through inquiry learning processes, students have better opportunities to engage with phenomena, develop inquiry skills and scientific reasoning, understand the meaning of doing and talking science, develop epistemological awareness of the nature of science and develop positive attitudes towards science.

3 Opportunities and Constraints that IBST/L Has Generated for Science Education

IBSE can be approached as a strategy for educational process rather than a method of learning and teaching. It is focused on providing suitable scaffolds for meaningful learning. Nevertheless, a critical question in this context is what are the opportunities and constraints that IBST/L has generated for science education?

Firstly, inquiry-based learning offers opportunities to learners for achieving a *better understanding of science concepts, principles and phenomena.* Within an IBST/L context, learners are offered *experiences* in which they can develop an understanding of science concepts and generally connect concepts and ideas with phenomena experienced in everyday life. Meta-analysis conducted by Schroeder et al., (2007) has shown that inquiry strategies demonstrated a statistically significant positive influence on students' achievements and learning when compared with the traditional teaching methods used in instruction of the control groups (Schroeder et al., 2007). They defined inquiry strategies as: *teachers use student-centered instruction that is less step-by-step and teacher-directed than traditional instruction; students answer scientific research questions by analyzing data (e.g., using guided or facilitated inquiry activities, laboratory inquiries)* (Schroeder et al., 2007; p. 1446).

Another opportunity that IBSL has generated for science learning is the development of general *inquiry skills* (Edelson, Gordin, & Pea, 1999) and *scientific practices* (NRC, 2012), such as posing and refining investigation questions, planning and managing investigations, gathering facts, exploring possibilities, conducting research, thinking through discoveries and analysing and communicating results (Alisinanoglu, Inan, Ozbey, & Usak, 2012).

When engaged with inquiry, especially with the aspect of *science explorations,* learners use and enhance their *cognitive skills*, such as analysing data and creating hypothesis which are essential competences for one's daily life (Alisinanoglu et al., 2012; Monteira & Jiménez-Aleixandre, 2016). Zoller (2011) introduces the concept of higher-order cognitive skills, referring to the same skills along with the ability to *transfer* those in different contexts. According to Zoller (2011), science learning should require the development of students' ability to be engaged in higher-order cognitive skills (HOCS), based on forms of inquiry such as *question asking, critical thinking, evaluative system thinking, decision-making and problem-solving capabilities in dealing with characteristically interdisciplinary everyday life*. An important element in this HOCS model presented by Zoller (2011) is the *transfer capability*, which is the capability of transferring different learning situations into real-life problem-solving contexts. Therefore, inquiry-based education proves vital especially with respect to achieving complex and comprehensive "higher-order" objectives such as understanding science principles, comprehending scientific inquiry and applying science knowledge to personal and societal issues (Anderson, 2002).

Moreover, IBSL offers opportunities to learners for developing scientific reasoning and gaining a better understanding of the nature of science, thus *developing epistemological awareness* (Chinn & Malhotra, 2002). According to recent science standards, the importance of learning to reason scientifically but also to comprehend the complex nature of scientific reasoning is stressed (Chinn & Malhotra, 2002).

Lastly, inquiry-based learning offers opportunities to learners for developing *positive attitudes towards science.* The inquiry approach allows students to connect classroom activities with their personal experiences and in this vein students are more motivated to learn. In the study of Rissing and Cogan (2009), significant gains were found in student attitudes when students participated in an inquiry laboratory (as cited in Gormally, Brickman, Hallar, & Armstrong, 2009). Also the findings of Gibson's study (1998) suggest that science programmes using an inquiry-based approach may help students with a high level of interest in science maintain that level of interest through their years in high school (Gibson 1998).

Empirical research provides evidence for the potential benefits that inquiry-based education might bring into students' cognitive, metacognitive and socioemotional domain, including (*1*) *cognitive achievements*, (*2*) *development of process and thinking skills*, (*3*) *development of attitudes towards science and provision of experiences with science* (Engeln, Mikelskis-Seifert, & Euler, 2014; Shymansky et al., 1983), (*4*) *development of scientific practices and inquiry skills* (Edelson et al., 1999; NRC, 2012) and (*5*) *development of epistemological awareness of how science operates* (Chinn & Malhotra, 2002) (see Table 1).

Large-scale evaluations have proven the effectiveness of inquiry learning over traditional modes of teaching (e.g. Linn, Lee, Tinker, Husic, & Chiu, 2006). Considering that inquiry experiences can provide valuable opportunities for students to improve their understanding of both science content and scientific practices, still the implementation of inquiry learning in classrooms presents a number of significant challenges (Edelson et al., 1999). In this respect, IBST/L has generated several *constraints* for science education, including (*1*) *the need for transformation of the national education standards* so as to be aligned with IBSL principles, (*2*) *the actual realization* of what constitutes "inquiry" in classroom terms, (*3*) *the degree of instructional support or guidance needed*, (*4*) *the difficulties that students* may encounter when being engaged with IBSL, (*5*) *the alignment of assessment methods* with IBSL principles and (*6*) *the teacher preparation and professional development* towards IBST (see Table 1). Even though science educators value opportunities that IBSL/T offers to learner, they often show reluctance in enacting IBST approaches in their teaching, as they consider those approaches as time-consuming leading to conflict with the requirement to *deliver curricula content* (Rocard et al., 2007). This demands an application of changes to curricula and methodologies by policy-makers. Moreover, the national education standards should encourage the active involvement of students in their own learning process, through group work and hands-on activities (NRC, 1996). Learning environments that support IBST/L should address certain characteristics of inquiry-based learning, such as considering and building upon students' prior knowledge (Hess & Trexler, 2005; Sewell, 2002), offering opportunities to students for supporting their conclusions with evidence and observations and prompting students to share and discuss their ideas with peers (Wolf & Fraser 2008).

A second constraint that has been encountered in efforts to promote IBSE is the recurring tendency to seek recipe-type representations of activity structures that can lead to classroom implementations. According to Chinn and Malhotra (2002), simple inquiry tasks have been introduced over the past two decades, as recipe-like approaches, in textbooks, trade books, educational software and websites of science activities (Chinn & Malhotra, 2002). Moreover, in recent years, this is fuelled by the trend in developing computer-based and online learning environments with a view to automate substantial aspects of scaffolding and guiding the student experience. This corresponds to typical guided and even transmissive way of teaching. Chinn and Malhotra (2002) argue that such *simple inquiry tasks* incorporate few if any features of authentic scientific inquiry. Procedures in most simple inquiry tasks are

Table 1 Opportunities and constraints that IBST/L has generated for science education

	Domain	Specific aspects
Opportunities	(i) Learners' cognition	Cognitive achievements: enhancement of cognitive skills
		Higher-order cognitive skill abilities – transfer capability
		Conceptual understanding of science concepts, principles and phenomena
	(ii) Learners' metacognition	Process skills
		Thinking skills/critical thinking
		Evaluative system thinking
		Decision-making
		Problem-solving capabilities
	(iii) Learners' socioemotional domain	Positive attitudes and motivation towards science
		Personal experiences with science
	(iv) Learners' scientific practices and inquiry skills	Posing and refining of investigation questions
		Planning and managing investigations
		Gathering facts and evidence
		Exploring possibilities
		Conducting research
		Thinking through discoveries
		Modelling
		Analysing and communicating results
		Argumentation
	(v) Learners' epistemological awareness	Scientific reasoning
		Scientific communication
		Understanding of the nature of science
		Understanding achievements of scientific discoveries
Constraints	(i) Alignment of national educational standards with IBSL/T principles	Curriculum updating and educational innovations
	(ii) Interpretation of what constitutes "inquiry" in classroom terms	Authentic inquiry vs recipe-like approaches
	(iii) Degree of instructional support or guidance needed in IBSL/T	Continuum between open and guided inquiry
	(iv) Difficulties that learners may encounter when being engaged with IBSL	Difficulties when conducting systematic scientific investigations due to the prerequisite of science content knowledge
		Scaffolding embedded in inquiry-based learning environments for overcoming learners' difficulties

(continued)

Table 1 (continued)

Domain	Specific aspects
(v) Alignment of assessment methods with IBSL/T principles	Adapting assessment methods used from deductive teaching approaches to inductive ones
	Summative vs formative assessment
(vi) Teacher preparation and professional development for IBSL/T	Changing the prevailing deductive teaching style is a highly challenging issue

straightforward, as students follow a short series of prescribed steps as in a recipe. However, in authentic research, procedures are complex and often require considerable ingenuity in their development.

Efforts to incorporate features of authentic scientific inquiry into classroom inquiry have led to much discussion about the degree of openness in inquiry learning which is the relative instructional support or guidance, even though inquiry learning has been widely recommended over the past years in science education (e.g. Blanchard et al., 2010; Chinn & Malhotra, 2002). While some studies found positive effects of open inquiry (e.g. Dochy, Segers, Van den Bossche, & Gijbels, 2003; Sadeh & Zion, 2009), others found negative effects in comparison to direct instruction (e.g. Chen & Klahr, 1999; Klahr & Nigam, 2004). For example, Shulman (1986) and Kirschner (1991, 1992) have independently reported problems on the content knowledge and the discussion processes that were impossible to carry on between experienced and novice learners, also at the epistemological level. Brown and Campione (1994) reported feelings of lost and frustration when students learn science in classrooms with pure discovery methods and minimal feedback from the teacher. Such situation, according to these researchers, caused students confusion which could lead to misconceptions. Alike, it was reported by Kirschner, Sweller and Clark et al. (2006) that "the minimal guidance during instruction is significantly less effective and efficient than guidance specifically designed to support the cognitive processing necessary for learning" (Kirschner et al., 2006, p. 76). These researchers explain this phenomenon referring to the human cognitive skills and constraints, such as cognitive load, epistemological differences between experts and novices, and human cognitive architecture. They also highlight that the advantage of offering guidance begins to recede only when learners have sufficiently high prior knowledge to provide "internal" guidance. Although these authors seem to misinterpret some of the strategies and methods in science education, generally they pointed out two reasonable problems: problems with students' prior content knowledge and the designing of accurate scaffolding by the teachers. A way to bridge this gap between direct instruction and open inquiry is *guided* inquiry, which combines the essence of open inquiry with instructional support (Furtak, 2006). It also indicates that this guidance should fade out while the educational process evolves.

According to Hmelo-Silver, Duncan, and Chinn, the teacher's support actually "plays a key role in facilitating the learning process" (2007, p. 100). The degree of direction or support offered by the teacher within each of the features of classroom inquiry identified above may vary along a continuum between *open* and *guided* inquiry (NRC, 2000)[1]. In practice, those distinctions and features of classroom inquiry are sometimes inadequately materialized by teachers and practitioners alike (Minner et al., 2010). This may come about from the lack of a shared understanding of the defining features of various instructional approaches that has hampered significant advancement in the research community on determining effects of distinct pedagogical practices.

As an aftereffect, constraint for science education is also the difficulties that students may encounter when *conducting systematic scientific investigations* (e.g. Krajcik et al., 1998; Schauble et al., 1995), since data gathering, analysis, interpretation and communication comprise challenging tasks. It is also essential that students have solid background knowledge on the topic that they are asked to inquiry. In particular, the formulation of research questions, the development of a research plan and the collection, analysis and interpretation of data are processes that require science content knowledge. When designing inquiry-based learning, it is a challenge to provide opportunities for learners to both develop and apply that scientific understanding. If students lack this knowledge and the opportunity to develop scientific epistemology, then they will be unable to complete meaningful investigations (Edelson et al., 1999). In open-ended inquiry learning environments, learners should also be able to organize and manage complex, extended activities. If they are not able to do so, students may face difficulties when being engaged in open-ended inquiry or achieve the potential of inquiry-based learning. Moreover, challenges may be confronted in practical implications, such as restrictions imposed by available resources and fixed schedules. Addressing the constraints of the learning environment is a critical consideration in design that must be considered alongside learning needs in the design of curriculum and technology (Edelson et al., 1999).

Research in the area of inquiry-based learning focuses on finding *adequate scaffolds* that help to prevent or overcome problems that students might confront while doing inquiry and that transform inquiry learning environments into effective and efficient learning situations (e.g. Hmelo-Silver et al., 2007). What is also still quite new and interesting in providing scaffolding for students is that computer environments can integrate cognitive scaffolds with the simulation (de Jong, 2006a; Linn, Bell, & Davis (2004); Quintana et al., 2004). Technological developments, such as computer simulations modelling a phenomenon or process (de Jong, 2006b) and hypermedia environments (Linn, Davis, & Bell, 2004), allow the effective implementation of inquiry learning.

A fourth constraint that IBSL/L has generated for science education is the *need for adapting the assessment methods* used from deductive teaching approaches to inductive ones. Looney (2011) argues that large-scale tests often do not reflect the

---

[1] Open and guided inquiry can be also defined as: "*Guided-inquiry* laboratories are experiments in which students follow experimental directions, gather data on certain specified variables, and through the analysis process establish relationships among the variables from their own data. *Open-inquiry* laboratories are experiments in which students design and perform their own procedures to investigate a question. Open-inquiry laboratories apply the relationships previously developed via guided-inquiry activities in a new setting or examine a new aspect of that relationship" (Chatterjee, Williamson, McCann, & Peck, 2009).

promoted development of higher-order skills such as problem-solving, reasoning and collaboration – which are key competences in IBSE. Moreover, the tradition of test-oriented and target-driven approaches of external testing leads to problems, including "teaching to the test", the detriment of the wider curriculum and motivational problems (Gardner, 2010). Especially high stakes connected to summative assessment often undermine innovative approaches to teaching (American Association for the Advancement of Science, 1998; Looney, 2011). Formative assessment has been seen as a means to achieve a better alignment between learning goals and assessment (Black & Wiliam, 1998), and in science education, it has received emphasis as a mechanism for scaffolding learning in science (e.g. Bell & Cowie, 2001). This is also supported by more recent European documents on formative assessment or a possible integration of formative and summative assessment, respectively (Centre for Educational Research and Innovation (CERI), 2008; Dolin & Evans, 2017; Looney, 2011; OECD, 2005).

Finally, a fifth constraint that IBSL/T has spawned for science education is *the need for new teacher professional development* (TPD) programs. Even though many good examples for inquiry-based learning in science education have been put forward by researchers, teacher educators and experienced teachers, changing the prevailing deductive teaching style is a highly challenging issue (Engeln et al., 2014). Teachers' professional competences are of crucial importance for keeping a proper balance between instruction and autonomous construction in the teaching and learning of science. According to Colburn (2000), even though "[…] there's no such thing as a teacher-proof curriculum, and there are lots of times when inquiry-based instruction is less advantageous than other methods. It's up to you to find the right mix of inquiry and non- inquiry methods that engage your students in the learning of science" (Colburn, 2000, p. 44).

In the European context, the confrontation of such constraints in IBST/L and ways of overcoming at least some of them demands the harmonization of educational standards and priorities. Changing the fragmented and colossal European educational systems requires extensive long-term efforts and involvement of all stakeholders at all levels. Those efforts involve the professional developments and continuous training and support of science teachers, the application of changes to curricula, the methodologies and assessment practices by policy-makers and the essential understanding but also support by parents on the need to have such changes. Also universities, business, local actors, informal science educators and civil society play a role in making science education more meaningful and linked to societal challenges, and research has to guide the change (Rocard et al., 2007).

4 Educational Policy Priorities for IBSE in Europe

Serious concerns are being raised about the status and the impact of science education and the decrease of students' interest for the corresponding key subjects (Engeln et al., 2014). In the European periphery, there has been a growing consensus, over the past decades, that the lack of substantial scientific literacy will hinder young people to become active and responsible citizens. The need for updating science-related studies and professions becomes significant for keeping up the pace of innovation and for reacting adequately to the economical, ecological and social challenges of a rapidly changing world.

The Lisbon summit emphasized on the need for European countries to act together to turn Europe into the most competitive knowledge-based world (Rocard et al., 2007). The summit recognized the need for action to foster a knowledge-based society and to promote education and training.

Reports by expert groups in the European context have identified the necessity of a renewed pedagogy that transforms the traditional mainly deductive teaching styles towards more appealing and cognitively activating forms of learning (Engeln et al., 2014). In this vein, inquiry-based science education has been identified as a method of choice for increasing students' interest and learning achievements in science (Rocard et al., 2007). Moreover, IBSE has been recognized to be appropriate to address the many societal and educational challenges (Rocard et al., 2007).

Reports by expert groups in the European context identify the "inquiry approach" as a "complex process of sense-making and constructing coherent conceptual models where students formulate questions, investigate to find answers, build new understandings, meanings and knowledge, communicate their learning to others and apply their learning productively in unfamiliar situation" (Beernaert et al., 2015, p. 68). In this approach students are engaged in authentic, problem-based learning activities, experimental procedures and "hands-on" activities, self-regulated learning sequences and discursive argumentation and communication with peers. The engagement of learners with IBSL develops a culture of scientific way of thinking and evidence-based reasoning skills for decision-making. It also develops competences for problem-solving and innovation, as well as analytical and critical thinking that are essential to empower citizens to have knowledge, skills and personal fulfilling for participating actively and responsively in an increased complex scientific and technological world (Beernaert et al., 2015).

Improving science teaching and encouraging more young people into the sciences have been key government objectives in Europe for years. This has led to the proposal and implementation of a variety of educational EU programs and projects for improving the quality of science education at school but also at the university level. In particular, the European Commission has financed, over the past years, many projects that focus on various ways to foster inquiry-based approaches in science education at all educational levels (e.g. PATHWAY, SCIENTIX, FIBONACCI, PRIMAS and others as reported in Beernaert et al., 2015). The fact that the European Commission (EC) has funded several research projects which are grounded in IBST/L reflects the establishment of IBST/L as an educational policy priority in Europe.

Despite the fact that science education community agrees on the fact that pedagogical practices based on inquiry-based methods are more effective, Rocard et al. (2007) in their report ascertain that the reality of classroom practices in most European countries is not aligned with the principles of IBSL/T. In fact, they report that in most European countries, science teaching methods are still essentially deductive. "The presentation of concepts and intellectual frameworks come first and are followed by the search for operational consequences, while experiments are mainly used as illustrations" (p. 15). Even though a change is under process in some countries towards more extensive use of inquiry-based methods, the majority still remains mainly deductive. In addition, they report that signs of an alarming decline in young people's interest for key science studies are still observable. They identify a connection between attitudes towards science and the way science is taught in classroom. Evidence to this indication comprise data from the 2005 Eurobarometer study on "Europeans, Science and Technology" which reports that only 15% of Europeans are satisfied with the quality of science classes in school. The report issued by the OECD "Evolution of Student Interest in Science and Technology Studies" identifies the crucial role of positive contacts with science at an early stage in the subsequent formation of attitudes towards science. The fact that some primary school teachers are requested to teach subjects in which they lack sufficient self-confidence and knowledge is also stressed in these reports. Teachers often choose a traditional "chalk and talk" approach with which they feel more comfortable and avoid inquiry-based methods that require them to have deeper integrated science understanding. The OECD report recommends the need to focus on teacher training in science, emphasizing in teaching approaches which concentrate on scientific concepts and methods rather than on retaining information merely. The High Level Group chaired by Prof. José Mariano Gago analyses problems found with science teaching in the report "Europe Needs More Scientists". They reach to similar conclusions that science subjects are often taught in a much too abstract way.

The above-mentioned evidence stress the need for identifying, integrating and disseminating good practices of IBSL/T. Teachers must remain the key players in the process of reform but need better support that complements professional training and stimulates confidence and motivation (Rocard et al., 2007). The adoption of new teaching techniques for stimulating inquiry-based learning among young people is one of the core actions that should be taken at the European level for achieving a substantive impact on the way that science is taught (Rocard et al., 2007). This is essential, because involving students, in inquiry-oriented science learning, compromises visions for the development of essential skills and competences essential for endorsing Responsible Research and Innovation (RRI), for enhancing public understanding of scientific findings including the capabilities to discuss their benefits and consequences and generally for cultivating responsible citizenship for the twenty-first century.

5   Implications for Science Teacher Development

Science teachers are probed to play a significant role for planning and implementing successful inquiry-oriented science lessons and therefore for supporting students' inquiry in the classroom. This challenging task demands from teachers to acquire, develop and utilize the abilities which are necessary for addressing students' inquiry.

Therefore, effective inquiry-based science teaching requires well-prepared and skilled teachers, who are aware of the essential characteristics and principles of inquiry learning and teaching, who can act as facilitators in their students' learning and who are ready to adapt and also develop inquiry-based teaching sequences in their everyday teaching. The essential preparation and expertise that teachers might need in this vein can be offered through well-designed professional development programs and pre-service preparation services.

A subsequent question that arises is: How to encourage sufficient and effective inquiry-based science teaching of good quality? First, it is important that teachers identify and overcome the potential barriers and obstacles while acquiring an inquiry approach to teaching science. Those barriers might derive from teachers' beliefs and values related to students, teaching and the objectives of education in general but also their understanding regarding the nature of science (Anderson, 2002). Teachers willing to embrace new approaches to education face dilemmas, which may rely on their beliefs and values and which can be clustered in three dimensions: technical, political and cultural dimension. On the top of that, teachers need to learn how to teach constructively in an inquiry-based approach, acquire new assessment competences and put their selves into a new teaching role. Thus the task for preparing teachers for inquiry-based teaching approaches is a challenging one.

In an effort to address this need, pre-service and in-service teachers should be given opportunities for becoming familiar with various inquiry-based approaches in science teaching and learning and also for participating in initial teacher preparation, induction programmes and professional development activities. For providing the science teacher education community with new knowledge, tools and strategies for IBST, it is vital to illustrate connections between knowledge deriving from the science education research community (scientific knowledge), from the science teachers' community (teaching practices) and from the educational innovation. It seems to be crucial to communicate with in-service teachers on the same epistemological level.

Inquiry may have a long history, and its definition may be still under discussion, likewise for inquiry-based science teaching. In our notion it is of greater emphasis to embrace this diversity and explore ways of achieving inquiry-based teaching of high quality, which presupposes well-trained and well-informed teachers and practitioners. Teacher development is progressive, since it leads to a better uptake of learners' needs, interests and competencies regarding the content and methods to be learned. Critical factors towards this progression constitute the nature of teachers' collaboration within the school and the uptake of in-service professional development programmes (Grangeat 2016).

6   Implications for Policy-Making for Science Education Reform

The lack of precise definitions of inquiry-based science education was first illustrated in a large-scale meta-analysis of the research literature by Anderson, in 1983. The same situation persists nowadays, in that inquiry teaching is described in different ways by different researchers, or the researcher may choose to use a different term for an approach that others would readily identify with the inquiry label (Anderson, 2002). Part of this lack of clarity is understandable, indeed we would argue unavoidable. It emerges from the fact that inquiry is a teaching-learning framework, a set of interconnected principles that need to be interpreted and contextualized in any effort to adopt them in the design of classroom practice or in the interpretation of teacher-learner or learner-learner interactions or even in the assessment of learning processes and outcomes.

However, another part of this haziness reflects an inherent systemic deficit in efforts to promote reform in science education. Educational policy-making tends to design reform as a set of administrative measures to be implemented in order to advance specific goals. In some cases, these goals are laudable, relating to teacher development, curriculum updating and educational innovations. The problem arises from the lack of mechanisms to connect reform with evolving communities of practice (Brown & Campione, 1994) and the need to engage professional communities in the actual rationale and design of dynamic reform initiatives. This is vital for creating learning communities that take advantage of and encourage distributed expertise within the community (Brown & Campione, 1994). Without such mechanisms of community engagement and dynamic evolution in the professional practices of teachers, it is only reasonable that oftentimes inertia will prevail, sustaining old established practices under the disguise of newly coined terminology, or that teams will treat the designed reform as an irrelevant administrative process which presents an opportunity to attract resources and promote discrepant agendas.

This phenomenon of systemic deficit in educational reform efforts becomes particularly acute when the design and monitoring of a reform initiative happens in agencies that are far removed from local educational systems or that are outside the structures of formal education failing to connect with established authorities with control over schools and teachers.

References

Alisinanoglu, F., Inan, H. Z., Ozbey, S., & Usak, M. (2012). Early childhood teacher candidates qualifications in science teaching. *Energy Education Science and Technology Part B, 4,* 373–390.

American Association for the Advancement of Science. (1998). *Blueprints for reform – Project 2061: Chapter 8: Assessment.* Retrieved from http://www.project2061.org/publications/bfr/online/blpintro.htm

Anderson, R. (2002). Reforming science teaching: What research says about inquiry. *Journal of Science Teacher Education, 13*(1), 1–12.

Arnold, J. C., Kremer, K., & Mayer, J. (2014). Understanding students' experiments—What kind of support do they need in inquiry tasks? *International Journal of Science Education, 36*(16), 2719–2749.

Ausubel, D. P. (1961). Learning by discovery: Rationale and mystique. *The Bulletin of the National Association of Secondary School Principals, 45*(269), 18–58.

Ausubel, D. P. (2012). *The acquisition and retention of knowledge: A cognitive view.* Dordrecht, The Netherlands: Springer Science & Business Media.

Baker, W. P., & Leyva, K. (2003). What variables affect solubility? *Science Activities, 40,* 23–26.

Barrow, L. H. (2006). A brief history of inquiry: From Dewey to standards. *Journal of Science Teacher Education, 17,* 265–278.

Beck, I. L. (1997). *Questioning the author: An approach for enhancing student engagement with text.* Order Department, International Reading Association, 800 Barksdale Road, PO Box 8139, Newark, DE 19714-8139.

Beernaert, Y., Constantinou, P. C., Deca, L., Grangeat, M., Karikorpi, M., Lazoudis, A., Casulleras, R. P., Welzel-Breuer, M. (2015). *Science education for responsible citizenship.* EU 26893, European Commission.

Bell, B., & Cowie, B. (2001). The characteristics of formative assessment in science education. *Science Education, 85*(5), 536–553.

Black, P., & Wiliam, D. (1998). Assessment and classroom learning. *Assessment in Education: Principles, Policy & Practice, 5*(1), 7–74.

Blanchard, M. R., Southerland, S. A., Osborne, J. W., Sampson, V. D., Annetta, L. A., & Granger, E. M. (2010). Is inquiry possible in light of accountability? A quantitative comparison of the relative effectiveness of guided inquiry and verification laboratory instruction. *Science Education, 94*(4), 577–616.

Blumenfeld, P. C., Soloway, E., Marx, R. W., Krajcik, J. S., Guzdial, M., & Palincsar, A. (1991). Motivating project-based learning: Sustaining the doing, supporting the learning. *Educational Psychologist, 26*(3–4), 369–398.

Brown, A. L., & Campione, J. C. (1994). Guided discovery in a community of learners. In K. McGilly (Ed.), *Classroom lessons: Integrating cognitive theory ad classroom practice* (pp. 229–270). Cambridge, MA: MIT Press.

Bruner, J. (1960). *The process of education.* Cambridge, MA: Harvard University Press.

Bybee, R. W. (2011). Scientific and engineering practices in K-12 classrooms: Understanding 'a framework for K-12 science education'. *Science Teacher, 78*(9), 34–40.

Cakir, M. (2008). Constructivist approaches to learning in science and their implication for science pedagogy: A literature review. *International Journal of Environmental and Science Education, 3*(4), 193–206.

Centre for Educational Research and Innovation (CERI). (2008). *Assessment for learning: Formative assessment.* Retrieved from http://www.oecd.org/site/educeri21st/40600533.pdf

Chatterjee, S., Williamson, V. M., McCann, K., & Peck, M. L. (2009). Surveying students' attitudes and perceptions toward guided-inquiry and open-inquiry laboratories. *Journal of Chemical Education, 86*(12), 1427.

Chen, Z., & Klahr, D. (1999). All other things being equal: Acquisition and transfer of the control of variables strategy. *Child Development, 70*(5), 1098–1120.

Chinn, C. A., & Malhotra, B. A. (2002). Epistemologically authentic inquiry in schools: A theoretical framework for evaluating inquiry tasks. *Science Education, 86*(2), 175–218.

Colburn, A. (2000). An inquiry primer. *Science Scope, 23*(6), 42–44.

Corbett, C. (2014). *Change in science teacher practice towards IBSE.* Doctoral dissertation, Dublin City University.

de Jong, T. (2006a). Computer simulations – Technological advances in inquiry learning. *Science, 312,* 532–533.

de Jong, T. (2006b). Scaffolds for computer simulation based scientific discovery learning. In J. Elen & R. E. Clark (Eds.), *Dealing with complexity in learning environments* (pp. 107–128). London: Elsevier Science Publishers.

Dewey, J. (1996). Essays. In L. Hickman (Ed.), *Collected work of John Dewey, 1882–1953: The electronic edition.* Charlottesville, VA: InteLex Corporation.

Dochy, F., Segers, M., Van den Bossche, P., & Gijbels, D. (2003). Effects of problem-based learning: A meta-analysis. *Learning and Instruction, 13*(5), 533–568.

Dolin, J., & Evans, R. (Eds.). (2017). *Transforming assessment: Through an interplay between practice, research and policy.* Cham, Switzerland: Springer.

Driver, R., Guesne, E., & Tiberghien, A. (1985). *Some features of children's ideas and their implications for teaching. Children's ideas in science.* Philadelphia, PA: Open University Press.

Driver, R., Newton, P., & Osborne, J. (2000). Establishing the norms of scientific argumentation in classrooms. *Science Education, 84*, 287–312.

Edelson, D. C., Gordin, D. N., & Pea, R. D. (1999). Addressing the challenges of inquiry-based learning through technology and curriculum design. *Journal of the Learning Sciences, 8*(3–4), 391–450.

Engeln, K., Mikelskis-Seifert, S., & Euler, M. (2014). Inquiry-based mathematics and science education across Europe: A synopsis of various approaches and their potentials. In *Topics and trends in current science education* (pp. 229–242). Dordrecht The Netherlands: Springer.

Erduran, S., Simon, S., & Osborne, J. (2004). TAPping into argumentation: Developments in the application of Toulmin's argument pattern for studying science discourse. *Science Education, 88*(6), 915–933.

Filipiak, E. (2011). *Z Wygotskim i Brunerem w tle: Słownik pojęć kluczowych*. Wydawnictwo Uniwersytetu Kazimierza Wielkiego.

Furtak, E. M. (2006). The problem with answers: An exploration of guided scientific inquiry teaching. *Science Education, 90*(3), 453–467.

Gardner, J. (2010). *Developing teacher assessment*. Maidenhead, UK: Open University Press.

Garrison, D. R., Anderson, T., & Archer, W. (1999). Critical inquiry in a text-based environment: Computer conferencing in higher education. *The Internet and Higher Education, 2*(2), 87–105.

Gibson, H. L. (1998). *Case studies of an inquiry-based science programs' impact on students' attitude towards science and interest in science careers.*

Gormally, C., Brickman, P., Hallar, B., & Armstrong, N. (2009). Effects of inquiry-based learning on students' science literacy skills and confidence. *International Journal for the Scholarship of Teaching and Learning, 3*(2), 16.

Grangeat, M. (2016). Dimensions and modalities of inquiry-based teaching: Understanding the variety of practices. *Education Inquiry, 7*(4), 29863.

Hancock, C., Kaput, J. J., & Goldsmith, L. T. (1992). Authentic inquiry with data: Critical barriers to classroom implementation. *Educational Psychologist, 27*, 337–364.

Harlen, W. (2013). *Assessment & inquiry-based science education: Issues in policy and practice.* Trieste, Italy: Global Network of Science Academies.

Haury, D. L. (1993). Teaching science through inquiry. *ERIC CSMEE Digest*, March. (ED 359 048).

Hayes, M. T. (2002). Elementary preservice teachers' struggles to define inquiry-based science teaching. *Journal of Science Teacher Education, 13*(2), 147–165.

Hess, A. J., & Trexler, C. J. (2005). Constructivist teaching: Developing constructivist approaches to the agricultural education class. *Agricultural Education Magazine, 77*, 12–13.

Hmelo-Silver, C. E., Duncan, R. G., & Chinn, C. A. (2007). Scaffolding and achievement in problem-based and inquiry learning: A response to Kirschner, Sweller, and Clark (2006). *Educational Psychologist, 42*(2), 99–107.

Hodson, D. (1993). Re-thinking old ways: Towards a more critical approach to practical work in school science. *Studies in Science Education, 22*(1), 85–142.

Jackson, S. L., Stratford, S. J., Krajcik, J., & Soloway, E. (1996). A learner-centered tool for students building models. *Communication of the ACM, 39*(4), 4849.

Jiménez-Aleixandre, M. P., Rodriguez, A. B., & Duschl, R. A. (2000). 'Doing the lesson' or 'doing science': Argument in high school genetics. *Science Education, 84*(6), 757–792.

Kirschner, P. A. (1991). *Practicals in higher science education*. Utrecht, The Netherlands: Lemma.

Kirschner, P. A. (1992). Epistemology, practical work and academic skills in science education. *Science & Education, 1*(3), 273–299.

Kirschner, P. A., Sweller, J., & Clark, R. E. (2006). Why minimal guidance during instruction does not work: An analysis of the failure of constructivist, discovery, problem-based, experiential, and inquiry-based teaching. *Educational Psychologist, 41*(2), 75–86.

Klahr, D., & Nigam, M. (2004). The equivalence of learning paths in early science instruction: Effects of direct instruction and discovery learning. *Psychological Science, 15*(10), 661–667.

Klus-Stańska. (2000). *Konstruowanie wiedzy w szkole*. Wyd. Uniwersytetu Warmińsko – Mazurskiego, Olsztyn.

Krajcik, J., Blumenfeld, P. C., Marx, R. W., Bass, K. M., Fredricks, J., & Soloway, E. (1998). Inquiry in project-based science classrooms: Initial attempts by middle school students. *Journal of the Learning Sciences, 7*(3–4), 313–350.

Kyza, E. A., Constantinou, C. P., & Spanoudis, G. (2011). Sixth graders' co-construction of explanations of a disturbance in an ecosystem: Exploring relationships between grouping, reflective scaffolding, and evidence-based explanations. *International Journal of Science Education, 33*(18), 2489–2525.

Linn, M. C., Bell, P., & Davis, E. A. (2004). Specific design principles: Elaborating the scaffolded knowledge integration framework. In M. Linn, E. A. Davis, & P. Bell (Eds.), *Internet environments for science education* (pp. 315–339). Mahwah, NJ: Lawrence Erlbaum Associates.

Linn, M. C., Bell, P., & Hsi, S. (1998). Using the internet to enhance student understanding of science: The knowledge integration environment. *Interactive Learning Environments, 6*(1–2), 4–38.

Linn, M. C., Davis, E. A., & Bell, P. (2004). Inquiry and technology. In *Internet environments for science education* (pp. 3–28). Mahwah, NJ: Lawrence Erlbaum Associates.

Linn, M. C., Lee, H.-S., Tinker, R., Husic, F., & Chiu, J. L. (2006). Teaching and assessing knowledge integration in science. *Science, 313*, 1049–1050.

Linn, M. C., Pea, R. D., & Songer, N. B. (1994). Can research on science learning and instruction inform standards for science education? *Journal of Science Education and Technology, 3*(1), 7–15.

Linn, M. C., Songer, N. B., & Eylon, B. S. (1996). Shifts and convergences in science learning and instruction. In *Handbook of educational psychology* (pp. 438–490). Riverside, NJ: Macmillan.

Looney, J. W. (2011). *Integrating formative and summative assessment: Progress toward a seamless system?* (OECD Education Working Papers No. 58). Retrieved from https://doi.org/10.1787/5kghx3kbl734-en

Mayer, R. (2004). Should there be a three-strikes rule against pure discovery learning? The case for guided methods of instruction. *American Psychologist, 59*(1), 14–19.

Mayer, R. E. (2002). Rote versus meaningful learning. *Theory Into Practice, 41*(4), 226–232.

Michalak, R. (2004). *Aktywizowanie ucznia w edukacji wczesnoszkolnej,* Wyd. Naukowe UAM, Poznań.

Minner, D. D., Levy, A. J., & Century, J. (2010). Inquiry-based science instruction—What is it and does it matter? Results from a research synthesis years 1984 to 2002. *Journal of Research in Science Teaching, 47*(4), 474–496.

Monteira, S. F., & Jiménez-Aleixandre, M. P. (2016). The practice of using evidence in kindergarten: The role of purposeful observation. *Journal of Research in Science Teaching, 53*(8), 1232–1258.

National Research Council. (1996). *National science education standards.* Washington, DC: National Academy Press.

National Research Council. (2000). *Inquiry and the national science education standards.* Washington, DC: National Academy Press.

National Research Council. (2012). *A framework for K-12 science education: Practices, crosscutting concepts, and core ideas.* Washington, DC: The National Academies Press.

NGSS Lead States. (2013). *Next generation science standards: For states, by states.* Retrieved December 1, 2013, from http://www.nextgenscience.org/next-generation-science-standards

OECD. (2005). *Formative assessment: Improving learning in secondary classrooms.* Paris: OECD Publishing and Centre for Educational Research and Innovation.

Osborne, J. (2006). Towards a science education for all: The role of ideas, evidence and argument. In *Proceedings of the ACER conference: Boosting science learning – What will it take?* Camberwell, VIC: ACER.

Osborne, J., Erduran, S., & Simon, S. (2004). Enhancing the quality of argumentation in school science. *Journal of Research in Science Teaching, 41*(10), 994–1020.

Pellegrino, J. W., & Hilton, M. L. (Eds.). (2013). *Education for life and work: Developing transferable knowledge and skills in the 21st century.* Washington, DC: National Academies Press.

Penner, D. E., Giles, N. D., Lehrer, R., & Schauble, L. (1997). Building functional models: Designing an elbow. *Journal of Research in Science Teaching, 34*, 125–143.

Prince, M., & Felder, R. (2007). The many faces of inductive teaching and learning. *Journal of College Science Teaching, 36*, 14–20.

Quintana, C., Reiser, B. J., Davis, E. A., Krajcik, J., Fretz, E., Duncan, R. G., et al. (2004). A scaffolding design framework for software to support science inquiry. *The Journal of the Learning Sciences, 13*, 337–387.

Rissing, S. W., & Cogan, J. G. (2009). Can an inquiry approach improve college student learning in a teaching laboratory? *CBE-Life Sciences Education, 8*(1), 55–61.

Rocard, M., Csermely, P., Jorde, D., Lenzen, D., Walberg-Henriksson, H., & Hemmo, V. (2007). *Rocard report: "Science education now: A new pedagogy for the future of Europe"*. EU 22845, European Commission.

Saari, H., & Viiri, J. (2003). A research-based teaching sequence for teaching the concept of modelling to seventh-grade students. *International Journal of Science Education, 25*(11), 1333–1352.

Sadeh, I., & Zion, M. (2009). The development of dynamic inquiry performances within an open inquiry setting: A comparison to guided inquiry setting. *Journal of Research in Science Teaching, 46*(10), 1137–1160.

Schauble, L., Glaser, R., Duschl, R. A., Schulze, S., & John, J. (1995). Students' understanding of the objectives and procedures of experimentation in the science classroom. *The Journal of the Learning Sciences, 4*, 131–166.

Schroeder, C., Scott, T., Tolson, H., Huang, T., & Lee, Y. (2007). A meta-analysis of national research: Effects of teaching strategies on student achievement in science in the United States. *Journal of Research in Science Teaching, 44*(10), 1436–1460.

Schwab, J. J. (1962). The concept of the structure of a discipline. *The Educational Record, 43*, 197–205.

Schwarz, C. V., & White, B. Y. (2005). Metamodeling knowledge: Developing students' understanding of scientific modeling. *Cognition and Instruction, 23*(2), 165–205.

Sewell, A. (2002). Constructivism and student misconceptions: Why every teacher needs to know about them. *Australian Science Teachers Journal, 48*, 24–28.

Shulman, L. S. (1986). Those who understand: Knowledge growth in teaching. *Educational Researcher, 15*(2), 4–14.

Shymansky, J. A., Kyle, W. C., & Alport, J. M. (1983). The effects of new science curricula on student performance. *Journal of Research in Science Teaching, 20*(5), 387–404.

Tschannen-Moran, M., & Hoy, A. W. (2001). Teacher efficacy: Capturing an elusive construct. *Teaching and Teacher Education, 17*(7), 783–805.

Vygotsky, L. S. (1971). *The psychology of art (Scripta Technica, Inc., Trans.)*. Cambridge, MA/ London: MIT Press (Original work published 1925).

Wilensky, U., & Resnick, M. (1999). Thinking in levels: A dynamic systems approach to making sense of the world. *Journal of Science Education and Technology, 8*(1), 3–19.

Windschitl, M. (2002). Framing constructivism in practice as the negotiation of dilemmas: An analysis of the conceptual, pedagogical, cultural, and political challenges facing teachers. *Review of Educational Research, 72*, 131–175.

Wolf, S. J., & Fraser, B. J. (2008). Learning environment, attitudes and achievement among middle-school science students using inquiry-based laboratory activities. *Research in Science Education, 38*(3), 321–341.

Zoller, U. (2011). Science and technology education in the STES context in primary schools: What should it take? *Journal of Science Education and Technology, 20*(5), 444–453.

Part I
Promoting Student Inquiry in the Science Classroom

Science Inquiry as Part of Technological Design: A Case of School-Based Development in Norway

Berit Bungum

1 Introduction

The current emphasis on inquiry-based science education (IBSE) in science teaching stems from various sources and is subject to diverse interpretations in science education (see, e.g. Duschl & Grandy, 2008; Schwartz, Lederman, & Crawford, 2004). As described in Chapter "Introduction: What Is Inquiry-Based Science Teaching and Learning?" of this book, its purpose either relates to views on learning (how children learn) or to views on the subject content (what children should learn in or about science). The former relates to constructivist perspectives on learning in science education (see Driver, Asoko, Leach, Mortimer, & Scott, 1994) and to earlier movements of discovery learning (see, e.g. Kirschner, Sweller, & Clark, 2006). The latter concerns the importance of learning about methods of scientific inquiry and, more generally, the nature of science (NOS). This latter perspective will form the foundation of the ideas put forward and the case reported in this chapter. The focus will be on how inquiry, which is representative of how scientists work in modern society, can form part of science teaching through interdisciplinary approaches within technological contexts.

These ideas will be illustrated using the case of a lower secondary school in Norway, where all students in Grade 9 (age 14–15) spend 2 entire weeks on a project designing and constructing their individual model cars. The case shows how school-based development can lead to sustainable change in how school subjects are presented to learners. This approach requires the school to adopt new and untraditional ways of thinking, and their strategies for justifying their practice will be presented. The author of this chapter had no official role in the project or within the

B. Bungum (✉)
The Norwegian University of Science and Technology, Department of Teacher Education, Trondheim, Norway
e-mail: berit.bungum@ntnu.no

© Springer International Publishing AG, part of Springer Nature 2018
O. E. Tsivitanidou et al. (eds.), *Professional Development for Inquiry-Based Science Teaching and Learning*, Contributions from Science Education Research 5, https://doi.org/10.1007/978-3-319-91406-0_2

school but collaborated with one of the teachers to document the project as inspiration for other teachers and schools, as part of the S-TEAM project (see Bungum & Jørgensen, 2010).

The chapter concludes with a discussion of the more general implications with regard to teachers' professionalism, school's autonomy and standardisation of the curriculum.

2   Inquiry and the Nature of Modern Scientists' Work

Inquiry-based learning has the potential of providing students with authentic experiences of how scientists work (see, e.g. Brickhouse, 2008). In Chapter "Introduction: What Is Inquiry-Based Science Teaching and Learning?", *scientific inquiry* – as one meaning of the term "inquiry" in IBSE – was described as the diverse ways in which scientists work to generate and validate knowledge. The teaching project presented in this chapter will be discussed with regard to how it reflects these authentic science practices and also extends the purpose of school science.

Many studies have shown that the general image portrayed in science textbooks and science teaching regarding the nature of science differs substantially from views held by philosophers of science (e.g. Alters, 1997) and those held by scientists (e.g. Wong, Hodson, Kwan, & Yung, 2008). Furthermore, even if experts agree on very general characteristics concerning the NOS, it is also clear that science constitutes a variety of diverse, local practices within and across scientific disciplines (Rudolph, 2000).

Science is often described in terms of how it works as a *collective* enterprise in order to establish new general knowledge of the natural world by developing theories and models. The approach taken in this chapter is rather to consider the nature of *individual* scientists' work in modern society. Although science, on the whole, may progress by means of formulating research questions and hypothesising, experimenting and falsifying of hypotheses in order to establish very generic theories of the world, this is rarely how individual scientists conduct their work today. They seldom formulate research questions; more often they form part of a large research group involved in long-term, extensive programmes and contribute by solving highly specialised theoretical and practical problems. The aim and products of this problem-solving are usually not grand theories but rather refined instruments and methods, more efficient experimental techniques and mathematical models that are slightly more predictive for specific purposes. Modern science is to a high-degree combinations of what Bybee (2011) has described as *science practices* and *engineering practices* in an educational context. For example, while science investigations are systematic ways of collecting data in the field or laboratory, engineering investigations are conducted to gain data essential for specifying criteria or parameters and to test proposed designs. Like scientists, engineers must identify relevant variables, decide how they will be measured and collect data for analysis. Their investigations help them to identify the effectiveness, efficiency and durability of designs under different conditions.

Furthermore, if we look at where people with a degree in science work after having finished their education, we find many of them outside a purely academic field associated with traditional universities. They do not strive to formulate grand theories at all but rather work within what Ziman (1984) identified as *post-academic science*, meaning the enterprise where scientists work for the purpose of industrial innovations and technological development. Modern science forms part of new alliances between universities, research centres and private corporations, where typically the aims are the improvement and refinement of existing systems, products and technologies. Scientists' work in these enterprises is characterised by creative problem-solving and systematic inquiries for highly practical purposes, within interdisciplinary settings.

This is in contrast to the picture often created of the independent and curious scientist taking part in inquiries to establish new generic knowledge of natural phenomena for its own sake. The primary function of post-academic science inquiry is essentially instrumental, and it justifies its existence in society by producing knowledge with a strategic and economical value, though sometimes this may be potential rather than actual. Thus, science has evolved from a novel search for new insights for purely academic purposes to being a driving force in economic development.

This transformation of science has important implications for the modernisation of science education. In order for students to experience what it means to work with science in modern society, they should be given opportunities to engage in purposeful, creative activities, develop their innovative skills and experience authentic science practices and engineering practices in meaningful contexts. These experiences as part of the curriculum should not primarily aim at learning content knowledge. A major study from Norway has documented that very little subject content knowledge from school science forms part of students' problem-solving in technology (see Bungum, Esjeholm, & Lysne, 2014). Rather, they apply and develop technological forms of knowledge. However, the combination of science and other subjects in creative technological contexts has a great potential for familiarising young people with what it means to work with science and technology in modern society, which is important for future employment as well as for citizenship. Substantial innovations in this direction require fundamental changes not only in how we look upon the aims and content of school science and other subjects. With regard to integration of subjects in cross-curricular approaches, Venville, Wallace, Rennie, and Malone (2002) have argued that integration, rather than being a combination of curriculum elements, is a particular ideological stance with roots in a view of knowledge that is worldly, experiential, contextual and organic. This stance is at odds with the disciplinary structure of schooling they describe as mechanistic, objective and framed within subjects. The school project described in this chapter is in line with the view of integration described by Venville et al. (2002) and also corresponds with how modern scientists and technologists combine knowledge in dynamic ways in order to reach solutions to problems.

3   School-Based Development and Teachers as Professionals

The project described in this chapter is a case of school-based development, characterised by bottom-up approaches aiming at innovations that affect the entire school through extensive teacher collaboration and development of local curricula (see, e.g. Ramberg, 2014). Governments and educational authorities often encourage innovative school development in their rhetoric. Yet, their regulations may in fact be an obstacle for innovations, since such innovations are not always in tune with standardisation and regulation of work in school. In a comprehensive case study for establishing an innovative school, Tubin (2009) identifies how institutional regulations need to be negotiated in order to realise and sustain innovative potential. The findings suggest three ways of handling regulation constraints: (1) setting an exception, (2) reallocation of resources and (3) adopting alternative standards. Setting an exception requires that all participants share a vision providing justification for the innovation, even if it operates at the edge of constraints set by school regulations in any given society. Reallocating resources, in terms of division of labour and co-ordination between teachers, is important since extraordinary resources directed to innovations are not likely to last long. Finally, standards for evaluating new and innovative practice must be adapted to the visions of the innovation. This can often – at least in principle – be done by reinterpreting given standards in formal curricula and regulations.

The influential traditions Tyack and Tobin (1994) describe as the "grammar of schooling" are an important barrier to innovation in schools. This term denotes the remarkably stable structures that organise school life, such as division of learning and activities into academic subjects, division of learners into groups according to age and timetables where every week follows the same pattern. These structures are passed on to new generations of teachers through "apprenticeship of observation" (Lortie, 2002) in many years as students. Patterns are hence reproduced by teachers and educational systems without explanation or reflection; it is deviations from these patterns that require justification.

Systematic studies have confirmed that teachers represent the most important factor for students' learning in schools (see Hattie, 2009), not only as individuals but also as representatives of a general school culture and of more specific traditions associated with specific school subjects (Siskin, 1994). Thus, sustainable change in schools requires that teachers are recognised as professionals and the driving force in school development. Professional practice has two main characteristics (see, e.g. Hoyle & John, 1995): a professional knowledge base shared by members of the profession and a professional autonomy associated with their professional practice.

It is paradoxical that in a time when major resources are committed to education and professional development programmes for teachers, educational systems are developing high levels of standardisation in teachers' work, driven by increasing requests for detailed reporting to school authorities on all levels, and standardised assessments locally, nationally and even internationally. This development points towards a bureaucratic control regime rather than professionalism for teachers. Skerrett and Hargreaves (2008) have, however, shown that this kind of standardisation has been counteractive in acknowledging diversity among students. They call for educational policy to involve fewer governmental regulations and less control and to be replaced by networking between schools, communities and highly qualified professionals in cultures of trust, cooperation and mutual responsibility.

Individual teachers and the cultures they are part of in terms of subjects, professional identities and educational thinking in the school and in society more generally are important for sustainable reforms which affect the entire school, not only single subjects (Black & Atkin, 1996). Substantial change in schools relies on and affects how one views students, subjects and the purpose of education in a broad sense. This means that teachers should be deeply involved in collaboration across subjects in order to achieve sustainable improvements that affect the entire school. An example of how this can be realised and maintained through school-based development is presented and discussed in the following section, through an example of a cross-curricular teaching project that involves new ways of thinking. The case illustrates how a school has managed to combat aspects of the grammar of schooling in their local curriculum development through constructive collaboration between teachers and integration of subjects in a joint project.

4   The Project "Wheels on Fire"

"Wheels on Fire!" is a cross-curricular teaching project in design and technology developed at Ruseløkka School in Oslo, Norway, and run every year for all students in Grade 9 (age 14–15). It lasts for 2 entire school weeks, and the students do not have any other subject lessons during this time. The framing of the project is slightly changed from year to avoid turning the innovation into pure routine for teachers. However, the main activity of the project is always that all students design and build a model car individually, made from plastic and run by an electric motor. The extraordinary aspect of the project is not the activity itself (many schools let students design and make model cars of one sort or another) but rather the time this school invests in the project, students' gains in terms of meaningful learning and motivation and how the time invested provides for a practical and creative approach to subject knowledge.

The following sections provide a description of how the project relates to the national curriculum in Norway, how it is organised at the school and what this entails for teachers and students. A detailed and practical teacher guide to the project is given by Bungum and Jørgensen (2010), together with video documentation on DVD available upon request from the authors.

4.1   Design and Technology in the Norwegian Curriculum

The project *Wheels on Fire* is framed within design and technology in the Norwegian curriculum. This subject area was formulated in the Norwegian national curriculum as a cross-curricular topic with a major curriculum reform in 2006. It involves the subjects science, mathematics and arts and crafts, for all students of Grades 1–10 (compulsory school grades). In science, the curriculum requires that students work with electronics in designing a product during Grades 8–10 (age 13–16), and in arts and crafts, "design" is a curricular area. The curriculum in mathematics states that students should be able to apply their mathematical knowledge in various contexts such as in design and technology. (The curriculum can be downloaded from Utdanningsdirektoratet, 2006.)

There is little systematic knowledge on how schools approach the cross-curricular topic, but a small-scale investigation has indicated that many schools do not put much effort into it (Dundas, 2011). The case of school-based development presented here is from a school that has succeeded in integrating subjects in constructive, cross-curricular, design and technology projects for their students. This serves as an example of how schools and teachers may develop and improve teaching within the frames of a national curriculum but grounded in the school's identity and the professional practice established by the team of teachers.

4.2   Organising the Project

Ruseløkka School has been running the project in Grade 9 annually for many years and involves approximately 10 teachers and 80–90 new students each year. The students work on the project during the whole timetable for 2 weeks, while teachers also keep up some scheduled teaching in other grades during the project period. Time resources are reallocated from the participating subjects, and subjects not represented in the project get "paid back" in terms of extra hours for teaching at other times of the year. It is clearly possible to run the project without using all teaching time during the project period. However, experiences from the project indicate that this particular organisation results in a more intense pace in students' work; the project is regarded by teachers as well as students as the main activity during these weeks, not as some extracurricular activity that adds to the "real" teaching of subjects.

Learning objectives for the project are formally the curriculum elements from the participating subjects. When the project is introduced, students are informed about the ways in which various subjects come into the project, for example, by using geometry in mathematics and work with electric circuits in science. However, teachers and school see the objectives in flexible and integrated ways and value students' experimentation, ability to make plans for their work, creative design skills, technical problem-solving and diligent work.

Teachers participating in the project have various subject backgrounds: arts and crafts, science, mathematics, social science and languages. The contribution from teachers of different subjects is a prerequisite for success in making subject content visible in the project. At the same time, teachers are challenged to identify how subject content can be made an integral component of the project as a whole. At Ruseløkka, teachers are accustomed to working in teams across subjects in all parts of their work, providing for this kind of collaboration within a cross-curricular project. It is, however, essential to have one dedicated teacher responsible for maintaining an overview of the preparation and implementation of all aspects of the project. This teacher is also in charge of the material and equipment to be used in the project (Norwegian schools do not have technicians for this purpose). Moreover, the school has put some effort into making sure all participating teachers are familiar with all practical aspects of the work students are undertaking, with regard to designing and building their individual model cars. This can only be achieved by going through the process oneself, and a practical workshop is, therefore, undertaken by all new participating teachers. This workshop is organised as an evening event and includes a group meal, which contributes to a good atmosphere and the formation of a community of practice (Wenger, 1998) among the teachers.

The school also invites external contributors to the project. These have been designers, car companies and representatives of other occupations, which contribute to connecting the students' work with the wider society. It is, however, essential that the school does not rely entirely on these external contributors in realising the project, since schools' partnerships with companies, universities, etc. easily fade over time. The responsibility and capability for running the project must therefore remain within the school and form part of their professional development. This has been an important policy at Ruseløkka School, where the design and technology projects form part of the school's identity.

4.3   Into the Classroom: How Students Work in the Project

For the students, the project "Wheels on Fire" represents a 2-week period where they work solely on the project, and do not have other lessons. An important aspect of the project is that students work in their usual classroom throughout the project, except during activities involving soldering the electric circuit. This gives continuity to students' work and the project does not affect timetables and resources for students in other grades. Equipment and tools used in the project are adapted to this situation and suitable for mobility and use in ordinary classrooms. Students design and create their model cars individually but cooperate in solving the tasks related to this. An example of how the car model might look is shown in Fig. 1.

The design and construction of model cars is placed within a context of entrepreneurship. Students play the role of designers engaged by a large car company which is concerned about the environment, user-friendliness and security of the vehicle. The students are invited to design a car that fulfils requirements with

Fig. 1 Example of the exterior and interior of a car model

regard to technical solutions, design and aesthetics. The students then work individually and at their own pace, but all go through the following phases (further details are given in Bungum & Jørgensen, 2010):

1. *Expressing ideas using mood boards*. The mood board is a visual expression of students' ideas with regard to the style of the car, the target user and some main features of design.
2. *Making a technical drawing*. The drawing shows all the parts of the car with detailed measures.
3. *Making a cardboard model*. The cardboard model is important in order to explore how the model car can be shaped in three dimensions. Sometimes students will have to go back to the technical drawing to make adjustments. In particular, students will have to consider techniques for joining parts.
4. *Shaping the model in plastic*. All parts are cut out using a knife and ruler and shaped by the use of portable line benders.
5. *Soldering the electrical circuit*. The circuit consists of a motor, a battery, a two way switch and wires. Some students also include light-emitting diodes for the model's lights and experiment as to how these should effectively be included in the circuit.
6. *Mounting wheels and driving mechanism*. The motor is connected to the wheels using a rubber band from a small pulley on the motor pin to a pulley wheel on one (or two) of the axles. Students experiment with how the placement and size of pulley wheels affect driving properties.
7. *Testing and improving*. This is an essential part of the project and students' learning. Most model cars do not run properly during the first trial, and systematic investigations are undertaken in order to solve problems and identify opportunities for improvement.

The project culminates in a great rally where students compete in speed racing, steady driving, uphill climbing and design. The competitions are organised by the students who finish their model cars early, thereby also experiencing rather complex organising skills. Pupils in lower grades are invited to watch the rally, making it a major event in which the school management also participates.

At all phases students' work is inspected by one of the teachers before they move on to a new phase. This ensures a certain level of quality and that all students succeed in creating a product that ultimately works. Parallel to the practical work, students undertake assignment work in the involved subjects. Assignments might be to calculate the area and cost of the plastic material required for the car model (mathematics) or to present the car model and how it works in a foreign language (English).

Students receive supervision individually or in small groups depending on their specific needs. Only a session on electricity and soldering is undertaken as a class. This lesson covers curricular content on electromagnetism and how the motor works, and places the knowledge students have acquired in earlier science lesson in the context of the practical task of connecting the motor to the battery and switch in the model car. The session is presented to all classes by a science teacher, who also guides the students in soldering the circuit. Students then take their completed circuit back to their classroom and continue with the work. When students encounter problems with their circuit, they can return to the science lab for assistance at set times. This arrangement gives students responsibility for planning their work in order to get the supervision they need, rather than following instruction from a teacher at all times. It also allows teachers to spend time supervising students where necessary, rather than instructing and controlling 30 students at a time.

4.4 Student Autonomy and Learning Potential

Observation in the school and communication with teachers and students indicate that students' motivation and engagement is exceptionally high during the project; even students who perform less well in more conventional lessons make a great effort with their model car. This may be due to the fact that students are allowed to create an individual product based on their own ideas and at their own pace.

As one would expect, students approach the task in very different ways. Some put great effort into the creative aesthetics of the car, or in constructing it to look authentic in all possible details. Others focus more on the driving properties and experiment with how transitions with pulleys can be made more effective, how weight should be distributed and how lights can be connected with a minimum loss of motor efficiency. Regardless of focus, all students face several technical problems in making the car run properly. This occurs because all cars are individually shaped by hand and have weaknesses that can only be reduced by finding the best compromise with other concerns regarding the construction. Improving the car model motivates the students to a high degree and gives room for problem-solving and inquiries driven by the students' desire to succeed with their product.

What do students learn from this, in terms of the content of the school subjects? In the case of science, it would clearly be naïve to believe that they discover scientific knowledge concerning friction and energy transfer by themselves, by struggling with making a model car work. Other teaching approaches justified by constructivism or principles of "discovery learning" in the 1970s and 1980s have been criticised for representing "a warmed-up version of old-style empiricism" (see Driver, 1983; Leach & Scott, 2003). The approaches advocated in this chapter could easily be subject to similar criticism. However, the situations provided by the project offer a *potential* for learning and for carrying out systematic experimentation in a motivating context that resembles how scientists and technologists work. To realise this potential, it is important that teachers have a good overview of the problems students may encounter; that they master a repertoire of relevant concepts, principles and techniques; that they set high expectations; and that they engage deeply with students during the entire process.

It is also clear that students learn different things in the project due to the freedom they have in how they approach the task. Differences in learning outcome are, however, always the case in education. In the project "Wheels on Fire", the school has accepted that students' learning outcomes indeed are different. Students are offered learning situations where they can develop their knowledge and skills in a range of ways depending on their focus, abilities and interests but are also put in situations where they gain experiences they would otherwise not have. Students' reflections on the project are reported in Bungum (2013), where one student, for example, describes the project in contrast to ordinary teaching where "you sit down and listen to the teacher, write down what is said, and then you go home!" Another student describes it as liberating to work on what she calls a "proper task". This is in line with how the teachers look upon the project as liberated from the logic of school subjects and how Venville et al. (2002) have described curriculum integration as a particular ideological stance. Also, girls and boys with minimum technical interest and low self-esteem were highly fascinated by having soldered their own electric circuit for the practical purpose of making their cars go. This may be of great importance for recruitment to science and technology, as it gives students experiences of what working within these fields in modern society might mean and the feeling of succeeding in creating a working technological product.

5   How Does "Wheels on Fire" Reflect Authentic Inquiry?

It can easily be argued that there are more effective ways to teach the content of the curriculum than through a cross-curricular project like "Wheels on Fire". However, in order to evaluate the learning potential of engaging with science in design and technology, one needs to reconsider the prominence of academic disciplines in compulsory education. Carlsen (1998) has suggested that we need a more socially and technologically situated perspective. The inquiries students undertake when constructing and improving their car model in the described project can be seen as an example of science in the situated practice of design and technology. In a situated perspective, science knowledge needs to be reconstructed and adjusted to the practical context in which it is to be applied (Layton, 1991). Most of the cars the students make do not run properly at the first attempt, and there is always room for improvement. During this process of adjusting the various parts of the construction, students make inquiries into science-related phenomena such as the effect of weight distribution, how wheel grip can be improved, how friction in the motor mechanism can be reduced and ways to connect the lights with the motor in the electric circuit. This process of improvement hence gives students opportunities to use systematic inquiries representative of the science and engineering practices that form part of professional scientists' work in post-academic science.

In order for the project to be successful in this regard, it is important that high-quality requirements for the products are set. Only then will the use of knowledge from science and mathematics become relevant. It is also important to avoid a focus entirely on the final, constructed product. Hence, documentation of the process should be undertaken during the project, including technical specifications where concepts from science and mathematics are explicitly addressed. The request for quality and documentation requires that students are given enough time to explore various solutions and to undertake systematic inquiry in order to succeed. This again calls for highly dedicated teachers that dare and are allowed to explore new fields of knowledge and new ways of teaching and a school that permits them to act as professionals in this regard.

6   Maintaining the School-Based Development

The project "Wheels on Fire" and other projects in design and technology at Ruseløkka School have been developed and run for many years, and they still win students' and teachers' enthusiasm; the innovation has achieved sustainability. Clearly this does not come without challenges. The main challenges have been related to internal resistance from some teachers, questioning the use of time and the prominence the project is given in the realisation of the curriculum, and the extra work involved. This mirrors the reluctance teachers often show towards inquiry-based approaches more generally, due to time constraints (see Chapter "Introduction: What Is Inquiry-Based Science Teaching and Learning?" of this book). Hence, the school's practice and priorities need to be justified and debated constantly. Strategies applied in their justifications resemble those identified by Tubin (2009) and described in the introduction of this chapter. The school *sets an exception* by making design and technology a key component of the profile and identity of the school. What makes it exceptional is not this cross-curricular domain as such but that the school takes it seriously enough to devote a substantial component of resources in terms of teaching time to a project like "Wheels on Fire". Systematic work has been done to encourage new teachers and teachers who show some resistance to take an active part in the project and allow their influence on how the project is undertaken. It is also important how the project combines different subjects and that the time spent on the project is sufficient for students to work with tasks within those subjects. This ensures an ownership of the project for everyone involved and ensures that knowledge remains and develops within the school as a community, not being dependent on external contributors or too few individual teachers. The manner of working in "Wheels on Fire" and other projects has thus gradually developed to form an important component of the school's identity and is promoted in a range of ways to pupils, parents, educational authorities and the local community. This means that fading out the project would generate negative reaction and loud protests from younger pupils (and their parents), who look forward to it for many years prior to Grade 9.

The school applies *alternative standards* when relating their focus on design and technology to more fundamental views on the importance of meaningful learning in creative contexts and students as active learners with autonomy and responsibility. The values and merits of the project are evaluated against the standards that reflect the school's explicit vision, which is also fully in accordance with the general part of the national curriculum stating that schools should develop students' creativity, innovative skills and belief in their own uniqueness. This part of the curriculum also encourages interdisciplinary approaches and learning in integrated and meaningful settings. This, in fact, might be difficult to combine with a focus on content knowledge in subjects, as the benefits of curriculum integration are difficult to measure in a disciplinary perspective (Venville et al., 2002).

In realising its vision, the school has made a *reallocation of resources* in terms of teacher resources and teaching time. During the project, a large proportion of time is spent on students' individual work rather than on whole-class structured instruction. This allows for students' experimentation and individual supervision to a much greater extent than when the majority of time is spent on whole-class instruction and where many students do not pay attention. The reallocation of working time for both teachers and students allows teachers to spend time on more constructive supervision adapted to individual students and the situation at hand.

It is quite clear that the three strategies setting an exception, reallocation of resources and adoption of alternative standards are closely linked. Setting an exception requires use of alternative standards and can only be fully realised through the reallocation of resources. Hence, all strategies need to be deeply rooted in a consistent school vision, shared by all teachers as well as the school management.

The shared vision is important in order to maintain the spirit at the school. The project, and similar projects undertaken in other grades, has the full support of the school management, and all teachers must participate actively. For sustainable cooperation between teachers, it is essential that all of them are challenged to contribute their knowledge and skills and that their contribution is taken seriously. This is important in making elements from subjects visible in the project; otherwise, the school could not justify the time spent on it. To be successful, a project like "Wheels on Fire" must be a way of working, not just something that adds to ordinary work.

The deep involvement of all teachers is also important in order to create a sense of ownership of the project for as many teachers as possible. The school involves external contributors, like car companies and graduate college students of design, but is never dependent on these contributors. The development and the continuation of the project have relied heavily on the one responsible teacher who has been in charge of it, but both this teacher and the school management have been conscious of the importance of distribution of responsibility and ownership. On a local level, the project is an example of how a professional community is engaged in the actual rationale and design of an innovative reform that contributes to improving young people's education, as called for in Chapter "Introduction: What Is Inquiry-Based Science Teaching and Learning?" of this book.

The success of the project and its maintenance hence seem to build on a fine balance between, on the one hand, giving skilled and creative teachers opportunities to realise their ideas as individuals and on the other hand providing for systematic creative work in cooperation across the entire staff of teachers. This has led to a sustainable deviation from the traditional "grammar of schooling" (Tyack & Tobin, 1994) and from the current standardisation of work in schools, and it also involves taking risks. It requires courage from educational authorities, school leaders and teachers to take these risks.

7   Conclusion: Autonomy for Development

The need for taking risks as described above links closely to autonomy for schools and teachers and how this can contribute to improved education for our children. As pointed out in Chapter "Introduction: What Is Inquiry-Based Science Teaching and Learning?", inquiry involves a degree of autonomy and responsibility for learners and teachers. It is in line with how inquiry-based learning activities are commonly described as authentic and problem-based activities that may not have a correct answer and where student autonomy is emphasised (see, e.g. Linn, Davis, & Bell, 2004). These aspects also characterise the project "Wheels on Fire" described in this chapter.

The issue of student autonomy has, however, more wide-reaching consequences for how we think about schools and teachers, since students' autonomy is not possible without professional autonomy for schools. Just as the school described as a case in this chapter allows for different learning outcomes for their students, school authorities must allow for differences in how schools approach the curriculum and teaching methods they employ. Fulfilling the potential inquiry-based learning or a cross-curricular project like "Wheels on Fire" provides requires highly professional teachers and for collegial as opposed to bureaucratic control over practice (Klette, 2000). Teachers need to be able and willing to develop and combine new ideas within and across subjects with their professional practice developed over many years. This calls for a "post-standardised" curriculum (Skerrett & Hargreaves, 2008) that allows for flexibility and school authorities who encourage innovation and creativity in education. In turn, this requires professional development programmes that are school-based, and not only allow for creativity and flexibility, but that *requires* schools and teachers to set their own goals and use their resources creatively on developing meaningful experiences for young people across traditional school subjects. "Wheels on Fire" as a case shows that this is realistic to achieve.

References

Alters, B. J. (1997). Whose nature of science? *Journal of Research in Science Teaching, 34*(1), 39–55.

Black, P., & Atkin, J. M. (1996). *Changing the subject: Innovations in science, maths and technology education*. London: Routledge.

Brickhouse, N. (2008). What is Inquiry? To whom should it be authentic? In R. Duschl & E. Grandy (Eds.), *Teaching scientific inquiry. Recommendations for research and implementation*. Rotterdam, The Netherlands/Taipei, Taiwan: Sense Publishers.

Bungum, B. (2013). Making it work: How students can experience authentic science inquiry in design and technology projects. In M. H. Hoveid & P. B. Gray (Eds.), *Inquiry in science education and science teacher education*. Trondheim, Norway: Akademika Forlag.

Bungum, B., Esjeholm, B.-T., & Lysne, D. A. (2014). Science and mathematics as part of practical projects in technology and design: An analysis of challenges in realising the curriculum in Norwegian schools. *NorDiNa, 10*(1), 3–15.

Bungum, B., & Jørgensen, E. C. (2010). *Wheels on fire! A practical guide for teachers*. Trondheim, Norway: NTNU/S-TEAM. Available online http://www.ntnu.no/fysikk/steam

Bybee, R. W. (2011). Scientific and engineering practices in K-12 classrooms: Understanding 'a framework for K-12 science education'. *Science Teacher, 78*(9), 34–40.

Carlsen, W. (1998). Engineering design in the classroom: Is it good science education or is it revolting? *Research in Science Education, 28*(1), 51–63. https://doi.org/10.1007/bf02461641

Driver, R. (1983). *The pupil as scientist?* Milton Keynes, UK: The Open University Press.

Driver, R., Asoko, H., Leach, J., Mortimer, E., & Scott, P. (1994). Constructing scientific knowledge in the classroom. *Educational Researcher, 23*(7), 5.

Dundas, A. A. (2011). *Hva skjedde med teknologi i skolen?* Master's thesis, The Norwegian University of Science and Technology, Trondheim.

Duschl, R., & Grandy, E. (2008). *Teaching scientific inquiry. Recommendations for research and implementation*. Rotterdam, The Netherlands/Taipei, Taiwan: Sense Publishers.

Hattie, J. A. C. (2009). *Visible learning. A synthesis of over 800 meta-analyses relating to achievement*. London: Routledge.

Hoyle, E., & John, P. (1995). *Professional knowledge and professional practice*. London: Continuum International Publishing Group.

Kirschner, P. A., Sweller, J., & Clark, R. E. (2006). Why minimal guidance during instruction does not work: An analysis of the failure of constructivist, discovery, problem-based, experiential, and inquiry-based teaching. *Educational Psychologist, 41*(2), 75–86. https://doi.org/10.1207/s15326985ep4102_1

Klette, K. (2000). Working-time blues: How Norwegian teachers experience restructuring in education. In C. Day, A. Fernandez, T. E. Hauge, & J. Møller (Eds.), *The life and work of teachers. International perspectives in changing times*. (1. London/New York: Falmer Press.

Layton, D. (1991). Science education and praxis: The relationship of school science to practical action. *Studies in Science Education, 19*(1), 43–79.

Leach, J., & Scott, P. (2003). Individual and sociocultural views of learning in science education. *Science & Education, 12*(1), 91–113.

Linn, M. C., Davis, E. A., & Bell, P. (2004). Inquiry and technology. In M. C. Linn, E. A. Davis, & P. Bell (Eds.), *Internet environments for science education* (pp. 2–27). Mahwah, NJ: Lawrence Erlbaum Associates.

Lortie, D. C. (2002). *Schoolteacher: A sociological study*. Chicago: University of Chicago Press.

Ramberg, M. R. (2014). What makes reform work?–School-based conditions as predictors of teachers' changing practice after a National Curriculum Reform. *International Education Studies, 7*(6), 46.

Rudolph, J. L. (2000). Reconsidering the 'nature of science' as a curriculum component. *Journal of Curriculum Studies, 32*(3), 403–419.

Schwartz, R., Lederman, N., & Crawford, B. (2004). Developing views of nature of science in an authentic context: An explicit approach to bridging the gap between nature of science and scientific inquiry. *Science Education, 88*(4), 610–645.

Siskin, L. S. (1994). *Realms of knowledge: Academic departments in secondary schools.* Washington, DC: Falmer Press.

Skerrett, A., & Hargreaves, A. (2008). Student diversity and secondary school change in a context of increasingly standardized reform. *American Educational Research Journal, 45*(4), 913–945.

Tubin, D. (2009). Planning an innovative school: How to reduce the likelihood of regression toward the mean. *Educational Management Administration and Leadership, 37*(3), 404–421.

Tyack, D., & Tobin, W. (1994). The "grammar" of schooling: Why has it been so hard to change? *American Educational Research Journal, 31*(3), 453.

Utdanningsdirektoratet. (2006). *Curricula for subjects in primary and secondary school.* Oslo, Norway: Utdanningsdirektoratet (The Norwegian Directorate for Education and Training). Available from www.udir.no

Venville, G., Wallace, J., Rennie, L., & Malone, J. (2002). Curriculum integration: Eroding the high ground of science as a school subject? *Studies in Science Education, 37*(1), 43–83.

Wenger, E. (1998). *Communities of practice: Learning, meaning, and identity.* Cambridge, UK: Cambridge University Press.

Wong, S. L., Hodson, D., Kwan, J., & Yung, B. H. W. (2008). Turning crisis into opportunity: Enhancing student-teachers' understanding of nature of science and scientific inquiry through a case study of the scientific research in severe acute respiratory syndrome. *International Journal of Science Education, 30*(11), 1417–1439.

Ziman, J. (1984). *An introduction to science studies. The philosophical and social aspects of science and technology.* Cambridge, MA: Cambridge University Press.

Promoting IBSE Using Living Organisms: Studying Snails in the Secondary Science Classroom

Eliza Rybska

1 Introduction

The idea that teaching science should be connected with practice and experience is not new. Eliot (1898) remarked that education should result in "the power of doing himself [sic] an endless variety of things which, uneducated, he could not do" (p. 323). Dewey (1938) emphasized the role of experience in education, and Kilpatrick (1918) valued the use of purposeful acts and the project method. Bandura (1977) related learning with direct experience through "trial and error"; he differentiated observational learning into learning by example and learning by percept.

IBST/L posits that science learning should include instances of examining the natural world. If science provides a way of observing, recognizing and describing the natural world, then students need to be introduced to scientific ways of interrogating evidence in order to develop their scientific thinking. They should also learn how to use this way of thinking as a means of generating credible knowledge about our world (DeBoer, 2000).

Science lessons in the classroom do not usually reflect the way science really works. For example, Klus-Stańska (2012) highlights the limitations of conventional teaching in Poland that involves mostly teacher-centred effort to transmit knowledge. Teachers tend to plan their lessons based on curriculum or textbooks, without taking into consideration students' pre-existing knowledge or the need to promote active learning. Often they teach the way they were taught – by dominating the discussion or even lecturing (e.g. Hernik, 2015; Kennedy, 1999; Oleson & Hora, 2014).

By contrast, inquiry-based science education (IBSE) has been proposed as a framework for conceptualizing the priorities and values of authentic science

E. Rybska (✉)
Department of Nature Education and Conservation, Adam Mickiewicz University, Poznań, Poland
e-mail: elizaryb@gmail.com

© Springer International Publishing AG, part of Springer Nature 2018   43
O. E. Tsivitanidou et al. (eds.), *Professional Development for Inquiry-Based Science Teaching and Learning*, Contributions from Science Education Research 5, https://doi.org/10.1007/978-3-319-91406-0_3

teaching and learning. IBSE includes several features, such as active pupil engagement in the learning process with emphasis on supporting knowledge claims with observations, experiences or complementary sources of credible evidence; tackling of authentic and problem-based learning activities; and development of the skills of systematic observation, questioning, planning and recording with a purpose to obtain credible evidence (Constantinou, Tsiivitanidou, & Rybska, 2018). Promoting IBST/L in biology presents an additional challenge; while the benefits of observing and investigating living organisms might be substantial, students and teachers might find it difficult to work with live creatures in the classroom (Tamir & Hamo, 1980).

1.1 Educational Reconstruction for IBST/L

In this paper, we investigate the extent to which hosting snails in the science classroom might serve as a productive context for teachers and students to engage with IBST/L. The educational reconstruction paradigm arose from the need to develop formative scaffolds to facilitate the evolution of students' alternative frameworks into scientific knowledge. As Kattman, Duit, Gropengießer and Komorek argue "since learning takes place in particular situational contexts, the science concepts and principles may not be presented in the abstract form but have to be put in certain contexts also" (1996, p. 2). The paradigm identifies and interrelates three research tasks: (a) clarification of science content, (b) investigation into students' perspectives and (c) analysis, design and evaluation of learning environments (Niebert & Gropengiesser, 2013).

The term reconstruction refers to the design of a process of knowledge formation "in order to make the science point of view understandable and meaningful to learners [...]" (Kattmann, Duit, Gropengießer, & Komorek, 1996, p. 2). Lijnse and Klaassenreal (2004) stress that simply acknowledging students' pre-existing knowledge is not adequate. If teaching aims to facilitate understanding, it seems necessary to "allow students ample freedom to use and make their constructions explicit, for example, by means of social interactions with the teacher and/or peers" (p. 159). At the same time, students need to be guided and supported along a process of knowledge construction. To respond to the need to facilitate student engagement and motivation, Lijnse and Klaassenreal (2004) propose a problem-posing approach to engaging students in a learning process of elaborating their existing conceptual knowledge, experiences and belief systems.

From this perspective, educational reconstruction provides a research-informed paradigm for promoting the major features of IBST/L including active pupil engagement in the learning process with an emphasis on supporting knowledge claims with observations, experiences or complementary sources of credible evidence, tackling authentic and problem-based learning activities and participation in collaborative group work, peer interaction, construction of discursive argumentation and communication with others (Constantinou, Tsiivitanidou, & Rybska, 2018).

1.2   Teaching-Learning Sequences (TLS)

TLS is "both an interventional research activity and a product, like a curriculum unit package, which includes well-researched teaching–learning activities, empirically adapted to student reasoning. Sometimes, teaching guidelines covering expected student reactions are also included" (Méheut & Psillos, 2004, p. 516). TLS can be a product that emerges out of an educational reconstruction effort (Méheut, 2004).

The educational reconstruction paradigm adds a synthetic perspective to designing TLS. Duit and co-workers (2012) provide four key questions for planning a lesson scenario: *Why? What? How? By what?* The *Why* questions refer to intentions, aims and objectives, e.g. why must students learn this topic? The *What* questions refer to science content, i.e. the concepts and epistemological objects that students are expected to master. The *How* questions relate to the anticipated teacher-student and student-student interactions, and the *By what* questions focus on any media to be used in class.

The intervention presented in this chapter consists of a topic-oriented TLS using snails, and thus it focuses on teaching and learning at a micro-level. Three steps were taken while planning the TLS (cf. Lijnse & Klaassen, 2004):

(i) Developing the teaching-learning scenario – on the basis of the common alternative conceptions about snails, the curriculum and a didactical analysis of the content to be taught. Using the personal experience of the author, the scenario was created as "theory-guided bricolage"[1] (Gravemeijer, 1994).

(ii) An interactive phase of scenario testing and refining, creating a possible didactical structure, testing a variety of hands-on activities (their order, goals, posing problems, driving questions) and differentiating similar activities for different age groups.

(iii) Empirical validation – performing the intervention and evaluating the outcomes using research tools (pre- and post-test).

---

[1] *Bricoleur* is a person who uses, as much as possible, those materials that are available. And "theory guided" indicates the way in which selections and adaptations of tasks and didactical sequence is guided by a domain-specific instruction theory.

1.2.1 The Pedagogic Affordances of Snails[2]

Students' interests in biology overlap with their personal interests. Prokop, Prokop, and Tunnicliffe (2007), in cross-age studies, were able to show that biology, as a school subject, is more endorsed by students when they learn about animals rather than plants.

Snails are a beautiful and easy-to-classify group of organisms. Land snails are especially common in Poland such as the Roman snail and the white-lipped snail. On the other hand, they have slimy bodies and thus may seem unpleasant. For some children, snails are even disgusting because of the mucus they have on their body (Davey et al., 1998; Randler, Hummel, & Prokop, 2012), which could be associated with low aesthetic value and negative attitudes (Prokop & Jančovičová, 2013).

The biology of snails is important. Land snails play a substantial role in forest decomposition processes and contribute to soil nitrification through their decaying bodies, shells and faeces. Their shells can form a source of calcium for other organisms. They also play an important role in many food chains, as herbivores and as organisms eaten by other animals. Snails can consume fungi and play a role in fungal spore dispersal. Some land snails serve as important indicators of biodiversity and play a role in monitoring of climate change (Ehrenfeld, 2010; Trautmann & Krasny, 1998; Wolters & Ekschmitt, 1997). Twenty-three gastropod species are listed in the Polish Red Data Book of Animals (Głowaciński & Nowacki, 2004). Students rarely consider the role that animals play in ecosystems, and thus they tend to perceive snails as useless or at least less useful than mammals. On the basis of such content knowledge, the aim of the presented study was to design the TLS that would allow to change student's attitude towards snails.

1.2.2 Curriculum

Due to the role snails play in ecosystems, gastropods are an important part of the Polish biology curriculum. Two learning objectives refer to the investigated topic directly. One describes a key concept: students should learn about the world of organisms, including describing, organizing and recognizing the organisms, showing the relationship between structure and function at different levels of life organization and presenting and explaining the interdependence between the organism and environment. For example, at the fourth stage (age 15–19), students should be

---

[2] The term affordances was proposed by Gibson (1979). The environment of every animal, including man, is made up of the affordances (offerings), such as, for example, something to eat or a place to meet other representatives of a species. These offers include knowledge – and this is due to the cognitive predisposition of the human phenotype. The affordance's value depends on two processes: (1) design of the offer: someone or something needs to do the calculations and carry out the process leading to the emergence of knowledge and (2) receiving offers, and more precisely, from the recipient, who, in order to use them, must have appropriate cognitive abilities (Błaszak, 2013). So in this chapter I claim that it is possible to add value (pedagogical value) to the affordance that snails create.

able to compare the structure and life functions of snails, bivalves and cephalopods, recognize typical representatives of these groups and show the biological importance of molluscs.

The second is a general goal that refers to responsible citizenship and states that students are expected to develop positive attitudes towards the environment including a broad understanding of the importance of protecting nature and a sense of respect towards all living beings.

In addition, the latest version of the Polish biology curriculum includes three goals that relate to inquiry: students are expected to be active in reasoning and processing arguments, improve their awareness of biological research methodology and able to search, evaluate and use information.

Students' Ideas About Snails Students often possess inaccurate ideas about snails. For example, students perceive snails as rather disgusting animals (e.g. Davey et al., 1998). Among interesting features that students tend to attribute to snails are that they can freely come out of their shell (Rybska & Sajkowska, 2012) or that the shell is their home and it is generally empty or filled with food (Rybska, Tunnicliffe, & Sajkowska, 2014). Snail movement is difficult to understand and describe – do they slide and move on the surface of mucus? Even snails' morphological structure is tricky for pupils. For example, locating the eyes can be difficult for students; whether snails possess a nose or not, students find it intriguing (Rybska & Sajkowska, 2012; Rybska, Sajkowska, & Tunnicliffe, 2015; Rybska, Tunnicliffe, & Sajkowska, 2014).

2 Methods

The following research questions were formulated:

1. Can children's interaction with live animals change their attitudes towards snails?
2. Are structured interactions more or less effective in terms of a positive influence on children's attitudes towards animals?

2.1 Structure of the Intervention

The designed TLS followed an inductive approach. Students are given space and time to carry out observations and to experiment with snails without hurting them. The intent is for the teacher to support students in constructing their own knowledge.

We adopted two main learning objectives: (a) enhancing students' familiarity with and care towards snails and (b) developing an evidence-based argument that snails are alive. The intervention was organized into four sections. In the first section, students focused on the morphological structure of the snail through observations. In particular, students, working in pairs, were asked to recognise parts of the snails' body, observe the eyes and record the behaviour. Students also compared live snails with empty shells. They were asked to describe their observations and solve a series of tasks about species recognition. Students had to count snails' antennae and check on which pair the eyes are located by moving a white stick in front possible locations. Also, students were asked to discuss common misconceptions that finally arise from a popular nursery rhyme about snails having no horns, only antennae. A short introduction of common features of snails was provided by the teacher.

The ensuing activities of the next three sections engaged students in examining the following features of snails:

Movement This activity involved group discussions about the possible role of mucus, predictions and observations of snail behaviour when put on flour, hypothesising about the influence of ground type on the speed of moving on glass and on sandpaper and observing muscle contraction on glass. The next task involved making observations of snails moving upside down on a sheet of paper and on a glass slide. The students were able to observe the movement of snail muscles in the foot and to discuss the strength of the snail's foot.

Anatomical Structure of a Snail This topic was addressed mostly via a short multimedia presentation and observations of snails while eating and carrying out other physiological activities. The presentation demonstrated different species of snails from around the world and also allowed students to test their ideas and predictions about snails. Issues addressed included how many teeth does a snail have? How does a snail differ from the human anatomical structure? What is interesting about the snails' internal anatomy? This part was done by the teacher, so a deductive way of reasoning was followed here.

Physiology and Behaviour In this part of the intervention, students attempted to answer questions such as: Do snails respond to stimuli? Can snails see? Can snails smell water? Can snails smell different substances such as lavender? In order to answer these questions, students had to carry out an experiment which involved investigating whether snails can smell. At first students were told that the smell sensors of snails are on the first pair of antennae. Next the students were shown how to carefully, and without touching the snail, use a stick to test the snail's response to a distinctive smell. The chosen substances were oils with a smell of pine, cloves, lavender, orange and sandalwood. Students had to investigate which scent the snails liked the most and which one was the most repellent. Then students discussed how the information about snails' favourite scents might be helpful for society, for example, in gardening activities. The last task for volunteer participants was to put a snail on their face and describe their experience to the other students. This part of the intervention again gave students an opportunity to act, to experience, to formulate questions, to hypothesise and to carry out experiments. Some activities were thought experiments, for example on

Fig. 1 A snail reacting to water and seeing from the bottom while moving on the glass (fot. E. Rybska)

discovering possible colours that can be seen by snails. Most of them were carried out practically by the students themselves working in groups. In this way, students could check whether a snail has an eye or a nose and where they are located (Fig. 1).

In addition, TLS was structured to enact inquiry as the "intentional process of diagnosing situations, formulating problems, critiquing experiments and distinguishing alternatives, planning investigations, researching conjectures, searching for information, constructing models, debating with peers using evidence and representations, and forming coherent arguments" (Linn, Davis, & Bell, 2004).

3   Data Collection

Designing a Survey Part A. Survey questionnaires were designed on the basis of the literature review, one focusing on students' conceptual understanding of snails and ecology and the second one as an adaptation of record, a Likert-type inventory to record students' attitudes about snails. The methods used in developing, validating and administering the instruments were consistent with the work of Stephen Kellert (1996). The inventory items can also be seen in Kellert's attitudinal scales. The factor structure is based on the work done by Barney, Mintzes, and Yen (2005) for attitudes towards dolphins and is outlined in Table 1. This served as a pilot study on a group of 30 students.

Part B. The modified version of questionnaire was given to the students before and after the intervention. Data were categorized, and by discussing with others not involved in research academic, the second categorization was made in order to make sure that the answers will be assigned to exactly one category (dialogue act) and that these categories are distinct. In order to check the differences between the dependent groups, a non-parametric U-test was used.

Table 1 Factor structure of the adopted snail's attitudes inventory

1	**Humanistic attitudes**
	It is impossible to like snails
	I'm not interested in anything that is associated with snails
	I like snails
	Snails have the same feelings as people
	I do not like the sound that is created when you accidentally step on a snail
	I like to watch snails
	I would put a snail on my face
	Open question: If you meet someone who knows everything about snails, what question would you ask?
2	**Utilitarian attitudes**
	Humans can benefit from the protection of snails
	I would like to breed snails at home
	The most common snail is the Roman snail
	The Roman snail is edible.
	Open question:
	Give three descriptive words about snails
	How would you convince a colleague/friend to take action to protect snails?
3	**Ecoscientific attitudes**
	Snails are important in nature
	I would like to know more about snails
	I like it when after rainfall snails appear on the pavement
	I would like to examine if the snails like cheese
	Open questions: The most interesting snail behaviour is…
	What would happen if there were no snails?

Components eliminated from snail attitudes inventory (following the pilot implementation) were as follows: If snails became extinct, I would feel like I had lost a friend. People are more important than snails, so we have the right to kill them for our joy

3.1 Participants

Workshops were carried out at the university. The participants were students (between 16 and 18 years old) from 8 hight schools northen Poland. In a total of 115 students (from the second class of high school), 88 of them were girls and 27 were boys.[3] Teachers were asked to ensure that students of a range of abilities were participating in classes. Every student who participated in this research was previously taught about molluscs at school. None of them had seen a live snail in a lesson before.

---

[3] Such disproportion is quite typical for high school classes that are specialized in biology.

3.2   Procedure

After completing the pilot study, the questionnaire was constructed, and the instructions for the didactic intervention were finalized in consultation with another researcher. Every question and points of instruction were reviewed and (if needed) improved. The next step of the research involved conducting classes with living snails. The chosen snails belonged to *Cepaea nemoralis*.[4] Classes were provided by researchers according to the following instructions. In the beginning, participants completed the questionnaires. The two questionnaires contained (a) questions about attitudes and emotions towards snails and (b) questions about participants' knowledge of the animals. The main question was semi-structured with different photographs of snails available. Completing the questionnaires took no longer than 15 min.

Afterwards, students watched the short multimedia presentation about different species of snails. After the presentation every student was given a sheet of paper and an individual snail. The snail was placed on this clean sheet. Before starting the classes, students were warned about ethical issues and treating animals well. Finally, after all of these experiences, students had to complete the same questionnaire which they did at the beginning of the class.

4   Results

The first part of the questionnaire aimed at measuring students' general attitudes towards snails. Twelve different pictures of snails were presented to children. Some of the pictures presented snails in some activities – such as eating, sperm-cell exchange or laying eggs. Children were asked to say whether they liked a picture (1 point), had no feelings towards it (2 points) or disliked it (3 points).

4.1   Attitudinal Inventory

In most cases children's attitudes, as revealed by their responses towards particular pictures presenting snails, changed into more positive ones. The only exception was the picture presenting a snail (*Succinea putris*) attacked by a parasite (*Leucochloridium paradoxum*). Infected snails have an enormous tentacle which is colourful and can pulse, becoming longer and shorter (Fig. 2). Children didn't appreciate the presence of parasites and disliked the picture even after the intervention. The results are presented in Table 2 and in Fig. 3.

---

[4] It was the species that was available at the university at that time, but it could be any snail species.

Fig. 2 Photograph K. Land snail *Succinea putris* with parasite *Leucochloridium paradoxum* inside its left eye stalk. Phot. Marta Świtała

Table 2 Analysis of differences in general attitudes towards snails on 12 pictures before and after the intervention

Mark, picture and name of the snail	Median before	Median after	Z statistic	P-value
A. Roman snail on a leaf *	2	1	4.47*	<0.0001*
Slug: B. *Limax maximus* *	3	2	3.67*	<0.0005*
C. *Helicidea* snail mating *	2	1	3.50	<0.0005*
D. Red-orange slug *Arion rufus* *	3	2	4.76	<0.0001*
E. Striped snail *Cepaea hortensis* *	1.5	1	2.74	0.0060
F. Two specimens of *Limax maximus* mating *	3	2	3.80	<0.0005*
G. *Glaucus atlanticus* blue ocean slug *	1	1	3.50	<0.0005*
H. *Hypselodoris bullockii* sea slug with colourful gills *	2	1	3.71	<0.0005*
I. Snail laying eggs *	3	2	4.67	<0.0001*
J. *Achatina* on human hand *	2	1	4.37	<0.0001*
K. *Succinea putris* infected by a parasitic worm *Leucochloridium paradoxum*	2	2	1.76	0.0768
L. Snail eating mushroom *	3	2	4.85	<0.0001*

Data were analysed using the non-parametric U-test
Values marked * show statistically significant improvements (assumed significance level $\alpha = 0.05$)

Most of the photographs were viewed by students in a positive way after the intervention, except for the snail in picture K. This picture shows a snail *Succinea putris* infected by a parasite *Lecucochloridium paradoxum* (Fig. 3).

In next stage, students were given three semi-structured questions. In the first question, students were asked to give three words (adjectives) that describe snails.

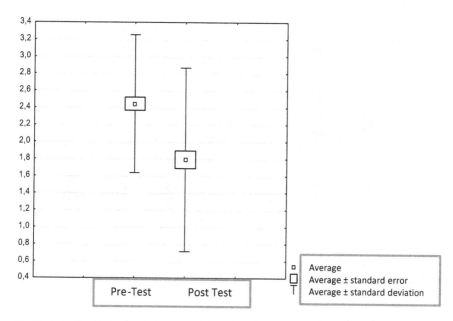

Fig. 3 Attitudinal inventory: Part A. Students' attitudes towards snails before and after the intervention, where *1* I like the snail on the picture, *2* I have no feelings towards the snail and *3* I don't like the snail in the picture (N = 115)

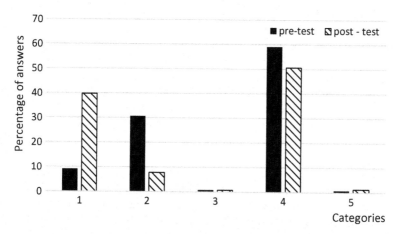

Fig. 4 Categorized descriptions of a snail given by students before and after the intervention. *1* indicating a positive attitude towards snails, *2* indicating a negative attitude towards snails, *3* neutral, *4* description of the animal, *5* no answer ($N_{pretest} = 144$, $N_{postest} = 156$)

Students' answers were categorized into four groups: 1, indicating a positive attitude towards snails; 2, indicating a negative attitude towards snails; 3, neutral; 4, description of the animal; and 0, for lack of an answer. Some students were placed

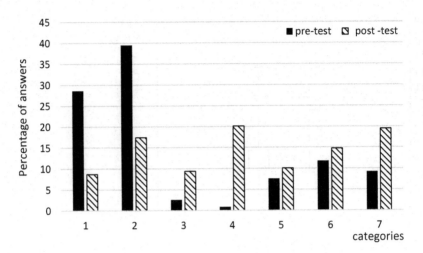

Fig. 5 Categorized answers to the question "What do you find interesting about snails?" Categories were *1* morphological trait, *2* anatomical feature, *3* senses, *4* behaviour, *5* physiology, *6* other features, and *7*, no answer ($N_{pretest} = 119$, $N_{posttest} = 149$)

in two or even three categories – according to the words recorded. The results are shown in Fig. 4.

The most frequently occurring words used by students before the intervention were slow, "slippery," "slimy" and "covered with mucus." In contrast, the most frequently recorded words after the intervention were "active," "always hungry" and "fast".

Although the pre-/post-test differences in this question were not statistically significant, some important findings emerged. Firstly far more descriptors were provided after the intervention than before (omission fraction with second and third adjectives on pretest = 0.78 and on post-test = 0.59). Secondly, only 12 students used a positive adjective about snails before (first category), while 62 did so afterwards. In the second category, before the intervention, 53 students showed a negative attitude towards snails, and after the intervention, the number reduced to 37 students out of 115.

Question number 2 was supposed to focus student attention on the snail as an interesting organism. Students had to answer the question "what is interesting about these animals?" As we might observe in the previous question, the omission fraction was much higher in the pretest (=0.36) than in the post-test (=0.18). Also the most common answer in this question before the intervention was that snails are able to leave their shell and find another one – which is not true. The most common answer in the post-test was that the most interesting in snails is that they react to stimuli. Again, the variety of answers was much wider after the intervention. Results are shown in Fig. 5.

Question number 3: "If you met someone who knew everything about snails, what would you ask?" Omission fraction on pre-test was 0.36 and on the post-test 0.46, which indicates that after the intervention, more students did not answer that question (Fig. 6).

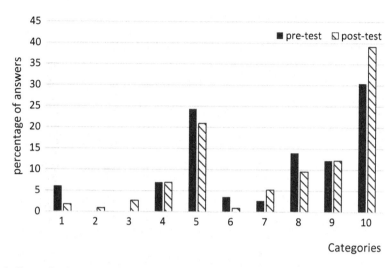

Fig. 6 Categorized answers to the question: "If you meet someone who knows everything about snails what would you asked?" Categories: *1* morphological trait, *2* anatomical feature, *3* senses, *4* behaviour, *5* physiology, *6* evolution, *7* curiosities, *8* personal question, *9* other question, *10* no answer ($N_{pre\&posttest}$ = 115)

Table 3 Examples of responses to each category to the question "If you meet someone who know everything about snails what would you asked?"

Category	Example of question
1 – Morphological trait	What is the largest/smallest snail? What are its tentacles for?
2 – Anatomical feature	What is inside the shell? Do they have a heart?
3 – Senses	Why do they react on some smells – like lavender?
4 – Behaviour	Can snails fight with each other? How do they communicate? Is it possible for a snail to fall in love with someone (human)? Do they like to imitate each other's behaviour? Do they feel sexual attraction? Which snail is the most dangerous?
5 – Physiology	How is poison produced in their body? Do you feel pain? Can a snail see colours? How long do they live? Could they survive in extreme conditions?
6 – Evolution	How did they evolve? What was the first snail on earth, and how did it look like?
7 – Curiosities	Where can I find the most interesting specimens? What is the maximum weight that they are able to hold? Can a snail walking on a knife edge hurt itself? What is interesting in snails?
8 – Personal question	Why would anybody spend his/her whole life investigating snails? What is the most interesting thing about snails? Why did you dedicate your life to them? What is his/her favourite snail?
9 – Other question	Can your research on snails help future generations? Are snails tasty? How to breed snails?

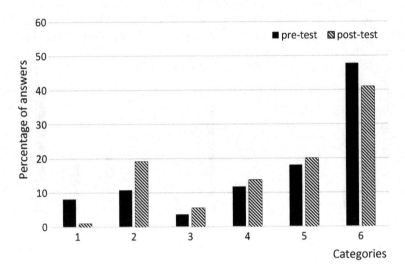

Fig. 7 Categorized answers to the question 4: "What would happen if there were no snails in the ecosystem?" Categories: *1* more plants would stay, *2* food deprivation for other animals, *3* reduction of biodiversity, *4* imbalance of ecosystems, *5* other effects, *6* no answer provided ($N_{pretest} = 111$, $N_{posttest} = 110$)

Table 4 Examples of responses in each category to the question: "What would happen if there were no snails in the ecosystem?"

Category	Example of response
1 – Related to plant	"My mom would have more plants in her garden"; "there would be more lettuce left"
2 – Related to animals	Ecological disaster – some species of other animals eats them; Birds would have a problem with food. Hedgehogs would not have enough food to eat and would break up the food chain
3 – Related to biodiversity	Animal world would be poorer; reduce diversity of species on earth
4 – Related to ecosystem	Changes in the whole ecosystem would appear; imbalance of ecosystems
5 – Other	The world would be more harmful; the world would be boring; French would be sad; oceans would not be so colourful; it would not be possible to use them for the production of cosmetics

Examples of responses to each category are presented in Table 3.

Question number 4. "What would happen if there were no snails in the ecosystem?" Omission fraction was 0.56 on the pretest and 0.46 on the post-test. Students' answers we categorised into four biologically relevant categories and two additional (were 5 – other effect, not relevant from biological/ecological point of view, and 6 – no answer provided) (Fig. 7).

Examples of such answers belonging to a particular category are presented in Table 4.

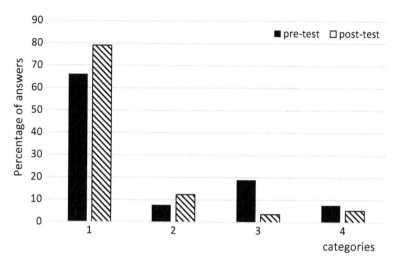

Fig. 8 Categorized answers to the question about the role of snails in ecosystems. Categories: *1* ecological/scientific, *2* humanistic, *3* utilitarian, *4* not significant (not connected to the topic) ($N_{pretest} = 53$, $N_{posttest} = 57$)

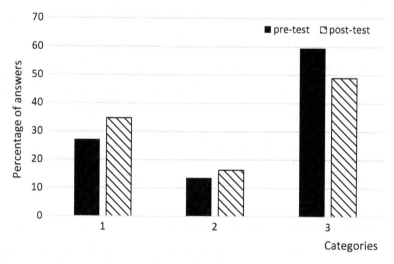

Fig. 9 Question 5. A person may benefit from the protection of snails, because.... Categories: *1* ecological/scientific, *2* humanistic, and *3* utilitarian ($N_{pretest} = 37$, $N_{posttest} = 49$)

Question number 5 was designed to check students' understanding of the role that snails play in ecosystems and their attitudes towards these animals. Students were supposed to complete four sentences. (a) Snails are important in nature, because...; (b) a person may benefit from the protection of snails, because...; (c) the most famous snail is...; and (d) the most interesting behaviour of snails is....

In the answers given for the first sentence – "snails are important in nature, because...." – there were four categories into which students' answers were

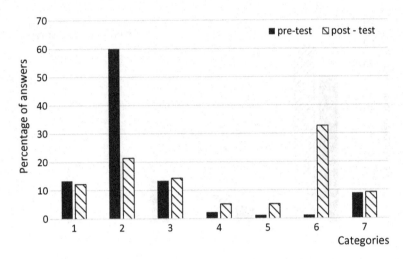

Fig. 10 Categorized answers to question 5d: What is the most interesting snails' behaviour? Categories: *1* reproduction, *2* possibility of hiding themselves into the shell or retracting their antennae, *3* movement, *4* eating, *5* mucus production, *6* reaction to stimulus, and *7* others ($N_{pretest} = 90$, $N_{posttest} = 99$)

categorized (Fig. 7). These are 1, ecological/scientific; 2, humanistic; 3, utilitarian; and 4, not significant (not connected to the topic). Graphical representation of data received for this task is shown in Fig. 8.

Second sentence: A person may benefit from the protection of snails, because….

Students' answers were categorized into three categories (also similar to Kellert scale), which were 1, ecological/scientific; 2, humanistic; and 3, utilitarian. The results are shown in (Fig. 9).

In question (5c), the most common answer was Roman snail – 90%. And on the (d) sentence after the intervention, the most common answer was respond to stimuli.

In question (5d) – The most interesting behaviour of snails is… – students' answers were categorized into seven categories on the basis of ecology, behaviour and physiological traits: 1, reproduction; 2, possibility of hiding themselves into the shell or retracting their antennae; 3, movement; 4, eating; 5, mucus production; 6, reaction to stimulus; and 7, others. The results are shown in (Fig. 10).

In question number 6, students were supposed to evaluate their attitude towards snails by providing an answer which is a statement on whether they do or not agree with the following sentences (in a scale of 1–5 where 1 is completely agree, 2 rather agree 3 neutral, 4 rather disagree and 5 completely disagree). In Table 5 are the results of research for this question.

The only two sentences that did not show a statistically significant change pre/post were as follows: (g) I like when after the rain the snails appear on the pavement, and (h) I do not like the sound that is created when you accidentally step on a snail. Only 17 students on the pretest completely agree that they liked it when snails appeared after the rain, and 11 rather agreed with it, while 38 completely disagreed, and 35 rather disagreed. On the post-test, 12 students completely agreed, and 12 rather

Table 5 Analysis of differences in general attitude towards snails on particular example of feeling towards snails between and after the intervention

Number of statement	Median before	Median after	Z statistic	p-value
6a. It is impossible to like snails *	3	4	3.41	<0.0005*
6b. I'm not interested in anything that is associated with snails *	3	4	4.66	<0.0001*
6c. I like snails *	3	3	4.53	<0.0001*
6d. I would like to breed snails at home *	5	4	5.40	<0.0001*
6e. I would like to know more about snails*	3	2	4.21	<0.0001*
6f. Snails have the same feelings as people*	4	3	4.95	<0.0001*
6g. I like when after the rain the snails appear on the pavement	4	4	1.32	0.1845
6h. I do not like the sound that is created when you accidentally step on a snail	1	1	0.15	0.8773
6i. I would like to see if snails like cheese	3	3	2,03	0.0421
6j I like to watch the snails *	3	2	5,53	<0.0001*
6k. I would put a snail on my face *	5	2.5	4,86	<0.0001*

Data were analysed using non-parametric U-test
The students were supposed to evaluate their attitudes on the scale (1 to 5) where 1 is completely agree, 2 rather agree 3 neutral, 4 rather disagree and 5 completely disagree (n = 115)
Values marked * are statistically significant

agreed, while 21 rather disagreed, and 35 completely disagreed. So, in summary, a number of students had negative attitudes towards snails appearing on a pathway after the rain decreased although this was not a statistically significant change.

On the pretest, 79 students agreed with the sentence "I do not like the sound that is created when you accidentally step on a snail" (65 completely and 14 rather agreed), while 18 disagreed (5 rather disagreed, and 13 completely disagreed). In the post-test, 73 students agreed with this sentence (64 completely agreed and 9 rather), while 23 students disagreed with it (7 rather disagreed and 16 completely disagreed).

5 Discussion

An attitude can be defined as the tendency to think, feel or act positively or negatively towards objects in our environment (Eagly & Chaiken, 1993; Petty, Haugtvedt, & Smith, 1995). It has been shown that knowledge of animals may influence children's beliefs and behaviour towards them (Thompson & Mintzes, 2002). Being a responsible citizen means being able to understand crucial elements of scientific literacy. That level of literacy also involves environmental insight, and typically it might be seen that environmental education (EE) programmes include activities aimed at encouraging pro-environmental attitudes and behaviours. Such programmes seek to educate "environmentally responsible citizens" with a view towards conserving natural resources and preserving and protecting the diversity of life forms in a variety of habitats (Barney et al., 2005). It was reported in research conducted by Thompson and Mintzes (2002) that knowledge of animals may influence children's beliefs and behaviour towards them. What's more, enhancing environmental knowledge may result in more positive pro-environmental attitudes (Hsu & Roth,1996). In the presented research, it might be noticed that before and after intervention, both knowledge and attitudes towards snails have changed. By interacting with living organisms, children became familiar with them as animated objects that react to stimuli and that are at some point like humans. It can be assumed from presented research that children's interaction with live animals change their attitudes toward snails into positive one, and that structured sequence of didactical situations that were designed on the basis of curriculum, recognition of student's conceptions about snails and planed interaction with animals helped them in this change. Especially all of the students have already been familiar with the content of snails from regular school classes.

On the other hand, much research attention has been dedicated to measuring factors that can influence behaviour. Variables that have been shown to correlate repeatedly with pro-environmental behaviour include verbal commitment, locus of control, attitude, personal responsibility and knowledge (Kollmuss & Agyeman, 2002). Although personality plays a very important role, among variables that are very strong are usually knowledge and attitudes (Barney et al., 2005). In the presented article, student's conceptions about snails were somehow shifted to a more scientific one (results are shown in Figs. 5, 7, 8, 9 and 10).

Poland is a country rich in small invertebrates such as snails. Although snails are among the animals that are not particularly liked or appreciated (Prokop & Tunnicliffe, 2010), it is possible to change attitudes about these animals as this study has demonstrated. Zoological issues are an important part of biology education. It is important that they be addressed in an attractive and meaningful way for students.

What is more important is that the interaction between animals and students in the classroom can have a positive impact on students' level of empathy for people and other animals (Daly & Morton, 2006). This study also found that students' attitude after interacting with snails changed.

Tamir and Hamo (1980) have shown that students are interested in learning with living organisms through direct observations and experiments. But at the same time, students were concerned about developing affection towards living organisms. Students were also more tolerant of the use of "harmful" animals or of "neutral" animals. What's more, younger students had more hesitation about the unrestricted use of plants and invertebrates or lower vertebrates, while older students were less likely to agree with sacrificing frogs and bats. From this point of view, such activities as presented above are neither harmful to animals nor connected to higher animals and so can be recommended for school use. It leads to the conclusion that structured interactions are more effective in terms of a positive influence on children's attitudes towards animals. Such claim is supported with presented results where structured intervention with living snails had a biggest influence on students attitude toward these animals.

Biologists who are exploring the various groups of animals and plants may be interested in shaping positive and pro-environmental attitudes among school students. Knowledge and positive emotions towards selected group of organisms, e.g. towards snails, can translate into more effective animal protection behaviours (Kellert, 1993). Millar and Osborne (1998) believe that the "science curriculum should provide sufficient scientific knowledge and understanding to enable students to read simple newspaper articles about science" (p. 9). The current Polish curriculum does not say anything about how to create positive attitudes towards nature or how involve emotions into educational process.

What is worth mentioning is that presented research can serve as a support for the idea of taking under consideration students' conceptions while planning lessons. It might be observed that Polish teachers usually plan their scenario on the basis of textbooks, their experience or sometimes the curriculum. In fact some of the teachers in Poland do not know the curriculum (Chęcińska, 2016; Hernik, 2015). As a result, even when some elements of IBST/L appear in it, teachers quite often are not aware of it. Reflective experienced teachers usually know some common students' conceptions and alternative conceptions, so they can prepare the scenario on this basis. But usually teachers do not plan their lesson on the basis of students' pre-existing knowledge.

The research presented here was done on the basis of participant responses rather than direct, long-term observations, which is a limitation; although at the end of intervention, there were students who really put snails on their faces – which is a clear behavioural aspect showing attitude towards snails.

6 Conclusions and Educational Implications

Structured interactions with live animals such as snails are useful and promising tools for changing student's attitudes. Such designed interventions might serve as "pedagogical affordances" for students.

For anybody who is involved with teaching about animals, the simple message would be as follows: Teach (about) life with life (living organisms)!

Proposed TSL as a sequence of didactical tasks can serve not only as a research activity but also as a product that might be used by a teacher while teaching about snails. One should encourage teachers to bring children to nature and provide opportunities for appreciation and respect – with adhering to principles of ethical treatment of animals.

References

Bandura, A. (1977). *Social learning theory* (pp. 305–316). Available at: http://www.esludwig.com/uploads/2/6/1/0/26105457/bandura_sociallearningtheory.pdf

Barney, E. C., Mintzes, J. J., & Yen, C. F. (2005). Assessing knowledge, attitudes, and behavior toward charismatic megafauna: The case of dolphins. *Journal of Environmental Education, 36*(2), 41–55.

Błaszak, M. (2013). *Ekotypy poznawcze człowieka. Przyczynek do kognitywistycznej teorii podmiotu.* Poznań, Poland: Bogucki Wydawnictwo Naukowe.

Chęcińska, K. (2016). *Praktyczne wykorzystanie metody eksperymentu na lekcjach przyrody w klasach 4–6 szkoły podstawowej.* Master thesis. Not published. In polish.

Constantinou, C. P., Tsivitanidou, O., & Rybska, E. (2018). *Introduction: What is inquiry-based science teaching and learning?* In Olia E. Tsivitanidou, Peter Gray, Eliza Rybska (eds) Professional development for inquiry-based science teaching and learning Springer Cham, Switzerland

Daly, B., & Morton, L. L. (2006). An investigation of human-animal interactions and empathy as related to pet preference, ownership, attachment, and attitudes in children. *Anthrozoös, 19*(2), 113–127.

Davey, G. C. L., McDonald, A. S., Hirisave, U., Prabhu, G. G., Iwawaki, S., Jim, C. I., et al. (1998). A cross-cultural study of animal fears. *Behaviour Research and Therapy, 36*(7–8), 735–750.

DeBoer, G. E. (2000). Scientific literacy: Another look at its historical and contemporary meanings and its relationship to science education reform. *Journal of Research in Science Teaching, 37*(6), 582–601.

Dewey, J. (1938, next 2007). *Experience and education.* Simon and Schuster. http://elibrary.kiu.ac.ug:8080/jspui/bitstream/1/1431/1/Experience%20and%20Education_0684838281-%20Dewey.pdf

Duit, R., Gropengießer, H., Kattmann, U., Komorek, M., & Parchmann, I. (2012). The model of educational reconstruction – A framework for improving teaching and learning science. In *Science education research and practice in Europe* (pp. 13–37). Rotterdam, the Netherlands: Sense Publishers.

Eagly, A. H., & Chaiken, S. (1993). *The psychology of attitudes.* Philadelphia: Harcourt Brace Jovanovich College Publishers.

Ehrenfeld, J. G. (2010). Ecosystem consequences of biological invasions. *Annual Review of Ecology, Evolution and Systematics, 41*, 59–80.

Eliot, C. (1898). *Educational reform.* New York: Century.

Gibson, J. J. (1979). *The ecological approach to visual perception.* Boston: Houghton Mifflin.

Głowaciński, Z., & Nowacki, J. (Eds.). (2004). *Polska czerwona księga zwierząt: bezkręgowce.* Kraków, Poland: Instytut Ochrony Przyrody PAN.

Gravemeijer, K. P. E. (1994). *Developing realistic mathematics education.* Utrecht, the Netherlands: CDBèta Press.

Hernik, K. (2015). *Polscy nauczyciele i dyrektorzy w Międzynarodowym Badaniu Nauczania i Uczenia się TALIS 2013.* Warszawa, Poland: Instytut Badań Edukacyjnych.

Hsu, S., & Roth, R. E. (1996). An assessment of environmental knowledge and attitudes held by community leaders in the Hualien area of Taiwan. *Journal of Environmental Education, 28*(1), 24–31.

Kattmann, U., Duit, R., Gropengießer, H., & Komorek, M. (1996, April). *Educational reconstruction–bringing together issues of scientific clarification and students' conceptions.* In Annual meeting of the national association of research in science teaching (NARST), St. Louis.

Kellert, S. R. (1993). Values and perceptions of invertebrates. *Conservation Biology, 7*(4), 845–855.

Kellert, S. R. (1996). *The value of life: Biological diversity and human society.* Washington, DC: Island Press.

Kennedy, M. M. (1999). The role of preservice teacher education. In *Teaching as the learning profession: Handbook of policy and practice* (pp. 54–85). San Francisco: Jossey-Bass.

Kilpatrick, W. (1918). The project method. *The Teachers College Record, 19*(4), 319–335.

Klus-Stańska, D. (2012). Wiedza, która zniewala–transmisyjne tradycje w szkolnej edukacji. *Forum Oświatowe, 1*(46), 21–40.

Koltmuss, A., & Agyeman, J. (2002). Mind the gap: Why do people act environmentally and what are the barriers to pro- environmental behavior? *Environmental Education Research, 8*, 239–260.

Lijnse, P., & Klaassen, K. (2004). Didactical structures as an outcome of research on teaching–learning sequences? *International Journal of Science Education, 26*(5), 537–554.

Linn, M. C., Davis, E. A., & Bell, P. (2004). Inquiry and technology. In *Internet environments for science education* (pp. 3–28). Mahwah, NJ: Lawrence Erlbaum Associates, Inc.

Méheut, M. (2004). Designing and validating two teaching–learning sequences about particle models. *International Journal of Science Education, 26*(5), 605–618.

Méheut, M., & Psillos, D. (2004). Teaching–learning sequences: Aims and tools for science education research. *International Journal of Science Education, 26*(5), 515–535.

Millar, R., & Osborne, J. (Eds.). (1998). *Beyond 2000: Science education for the future* (The report of a seminar series funded by the Nuffield Foundation). London: King's College London.

Niebert, K., & Gropengiesser, H. (2013). The model of educational reconstruction: A framework for the design of theory-based content specific interventions. The example of climate change. In *Educational design research–Part B: Illustrative cases* (pp. 511–531). Enschede, the Netherlands: SLO.

Oleson, A., & Hora, M. T. (2014). Teaching the way they were taught? Revisiting the sources of teaching knowledge and the role of prior experience in shaping faculty teaching practices. *Higher Education, 68*(1), 29–45.

Petty, R. E., Haugtvedt, C. P., & Smith, S. M. (1995). Elaboration as a determinant of attitude strength: Creating attitudes that are persistent, resistant, and predictive of behavior. In *Attitude strength: Antecedents and consequences* (Vol. 4, pp. 93–130). Mahwah, NJ: Erlbaum.

Prokop, P., & Jančovičová, J. (2013). Disgust sensitivity and gender differences: An initial test of the parental investment hypothesis. *Problems of Psychology in the 21st Century, 7*(7), 40–48.

Prokop, P., Prokop, M., & Tunnicliffe, S. D. (2007). Is biology boring? Student attitudes toward biology. *Journal of Biological Education, 42*(1), 36–39.

Prokop, P., & Tunnicliffe, S. D. (2010). Effects of having pets at home on children's attitudes toward popular and unpopular animals. *Anthrozoös, 23*(1), 21–35.

Randler, C., Hummel, E., & Prokop, P. (2012). Practical work at school reduces disgust and fear of unpopular animals. *Society & Animals, 20*(1), 61–74.

Rybska, E., & Sajkowska, Z. (2012). *Wiedza potoczna versus wiedza naukowa na temat ślimaków–możliwości i zaniedbania.* Badania w dydaktykach nauk przyrodniczych (Research in didactics of the sciences), pp. 126–130. In polish.

Rybska, E., Sajkowska, Z., & Tunnicliffe, S. D. (2015). What's in a shell? Children's concepts of snail anatomy. *Folia Malacologica, 23*(1), 78.

Rybska, E., Tunnicliffe, S. D., & Sajkowska, Z. A. (2014). Young children's ideas about snail internal anatomy. *Journal of Baltic Science Education, 13*(6), 828–838.

Tamir, P., & Hamo, A. (1980). Attitudes of secondary school students in Israel toward the use of living organisms in the study of biology. *International Journal for the Study of Animal Problems, 1*(5), 299–311.

Thompson, T. L., & Mintzes, J. J. (2002). Cognitive structure and the affective domain: On knowing and feeling in biology. *International Journal of Science Education, 24*(6), 645–660.

Trautmann, N. M., & Krasny, M. E. (1998). *Composting in the classroom.* Scientific inquiry for high school students. https://pdfs.semanticscholar.org/8fb7/0d9b0cdfa257ef4a65d7ea52c12e43db15c2.pdf

Wolters, V., & Ekschmitt, K. (1997). Gastropods, isopods, diplopods, and chilopods: Neglected groups of the decomposer food web. In *Fauna in soil ecosystems* (pp. 265–306). New York: Marcel Dekker, Inc.

Drama As a Learning Medium in Science Education

Ran Peleg, Anna-Lena Østern, Alex Strømme, and Ayelet Baram Tsabari

1 Introduction and Rationale

[Y]ou are working in one of the world's great art forms, the art form that lays bare human behaviour for us to examine and reflect on. – John O'Toole and Julie Dunn (2002, p. 2–3)

There is a growing concern in many European countries that the public status of science is falling (Osborne & Dillon, 2008). This is reflected in a decline in the interest of young people in the sciences and a lack of students who choose to specialize in them (Osborne, Simon, & Collins, 2003). This situation must be remedied, since a knowledge-based society derives its strength from future scientists, engineers and its science and technology workforce. In addition, in a democratic knowledge-based society, technological knowledge and skills are a must for engaging in the many science and science policy issues affecting modern society. Indeed, in a recent report on the state of science education in Europe, a call has been made for "innovative curricula and ways of organising the teaching of science that address the issue of low student motivation" (Osborne & Dillon, 2008, p.16). In addition, this report calls for curricula that enhance students' scientific literacy skills for future informed citizens and not just for future scientists.

R. Peleg (✉)
EdQuest – Creative Science Education, Rehovot, Israel

A.-L. Østern · A. Strømme
Department of teacher education, NTNU, Trondheim, Norway
e-mail: anna.l.ostern@plu.ntnu.no; alex.stromme@plu.ntnu.no

A. B. Tsabari
Faculty of Education in Science and Technology, Technion – Israel Institute of Technology, Haifa, Israel
e-mail: ayelet@technion.ac.il

This chapter aims to provide science teacher educators with theoretical and practical knowledge of how drama can serve as an inquiry-based teaching and learning tool in science education, in order to increase students' scientific literacy, engagement and motivation. It is our pre-understanding that drama, as an art form, introduces an added value in exploring scientific themes, through embodied learning. We begin with a theoretical framework for drama as an inquiry-based learning medium in science education from a sociocultural perspective. We then give some illustrative examples of how drama works in the science classroom. Finally, we address teacher training for using drama in the science classroom by addressing the skills required by teachers, including the use of formative assessment connected to clearly defined competencies to scaffold the learning process, and by providing two exemplary workshops.

2   Drama As an Inquiry-Based Teaching Method in Science Education

We base our discussion on the definition of inquiry defined in the first chapter, The framework in chapter "Introduction: What Is Inquiry-Based Science Teaching and Learning?" suggests five features for inquiry-based science education and learning:

(i) Learners being engaged by meaningful scientifically oriented questions
(ii) Learners giving priority to evidence, which allows them to develop and evaluate ideas that address scientific questions
(iii) Learners formulating knowledge claims and arguments from evidence in order to settle scientific questions
(iv) Learners evaluating their explanations in light of alternative explanations, particularly those reflecting scientific understanding
(v) Learners communicating and justifying their proposed explanations (Co-authors chapter "What Is Inquiry-Based Science Teaching and Learning?")

Drama techniques can support each of these features in the following manner as will be discussed in this chapter:

- Drama can serve as an entry point into conducting scientific inquiry (Peleg, Katchevich, et al., 2015; Peleg, Yayon, et al., 2017). This supports feature (i) above.
- To create a successful scientific drama, students must gather evidence thus supporting feature (ii).
- Drama can aid in processing complex information and promote critical thinking thus supporting features (iii) and (iv).
- In drama-based pedagogy, learners communicate their explanations to an audience and receive feedback thus supporting feature (v).

In this section we shall present the case that drama can serve as such an inquiry mode. We do not (and cannot) present prescribed recipes for inquiry-based science teaching using drama but rather outline the important and necessary elements for a successful teaching discourse. We will investigate what characterizes drama in science education in the classroom, how it works and how it may support inquiry.

Science education as a learning area in school context includes biology, chemistry, physics and perspectives of ethics and sustainable development in these areas.

2.1 What Characterizes Drama in Science Education?

Drama in general education contexts, according to O'Toole and Dunn (2013), is "both about exploring – discovering and creating – and about performing. Principally, especially in the primary years, it is about creating models – models of behaviour and action that can be practised and performed safely." They underline that through drama, active and realistic models of human behaviour can be created and experienced first-hand. Within the classroom, one can explore safely how people behave in other human contexts, all over the world and throughout history. O'Toole and Dunn further describe the drama classroom as a model space for exploring the world beyond in a safe way because we can always stop the pretend situation and walk away, unscathed but not untouched.

From the perspective of performance theory, drama can be described as action in role. The aesthetic doubling taking place consists of four doublings: doubling of time, space, role and story. You step into fiction, as well as in somebody else's shoes. The teacher of drama can be considered a dramaturge, who stages the classroom activities like a performance (Østern, 2014). Selander (2017) has suggested educational design theory as the post-Vygotskian mode of working, teaching and learning by design. Selander is not writing about drama, but he describes the consequences for education when the society gets more digitalized and the students design tasks in multimodal ways. A contemporary view of drama finds place both for devised performances based on research by the students and for explorative process drama sessions. Furthermore inspiration from performance in theatre makes the fusion of fiction and non-fiction an authentic way to devise documentary scenes from the history of science education and from the real controversies existing today within the field.

Drama in science education does not differ much from drama in general educational settings except for the themes handled. These themes have called for two general teaching strategies within the science classroom (Dorion, 2009). The first involves the simulation of social events. These can help students explore and understand the impact of science on society and do not differ much from the settings described by O'Toole and Dunn: they allow exploration of how people behave in other human contexts. Activities whereby the learners take on a human character will be referred to as "simulations" following McSharry and Jones (2000).

The second strategy employs mime and role play for presenting abstract physical phenomena and has been referred to as analogies by McSharry and Jones (2000). In such cases students no longer act as humans but rather as physical or natural entities such as molecules, animals, photons, etc. Students no longer experience other human contexts, yet they still create a mental model and experience this model first-hand in a multisensory manner. Examples of suitable topics may be found in McSharry and Jones (2000).

Another feature of drama in a science education setting that should be considered is the familiarity of the topics handled. While in many other subjects (such as history or languages), everyday experiences can form the starting point for drama, in science, the topics taught, especially those involving scientific phenomena, are often abstract and unfamiliar to the learner.

Drama activities in education are characterized by the following aspects: uses fiction, allows for mental models to be constructed and examined, allows for socio-cultural activities and for scaffolding in learning science, introduces imagination and creativity to the science classroom, allows for both narrative and logical-scientific thinking and explores situations and includes performance for the class-mates. All these aspects will be presented in following subchapters.

2.1.1 Drama Uses Fiction

Learners in drama are transported to an "as if mode," whereby this mode requires the students to act as physical entities such as molecules or as real people such as historical figures. This inquiry mode allows learners to construct and test explanations from a perspective other than their daily lives. Learners must be willing to accept and embrace the fictive world that will be constructed. Any drama activity should start with a dramatic contract in which participants agree to pretend and to willingly suspend their disbelief. In theatre, such a contract is established upon entering the theatre hall. In a classroom setting, such a contract must be "negotiated" or clarified with the learners. This will ensure that the participants' behaviour as characters will not "spill" into the real world and ensure a level of confidentiality and liberty. Students should experience a safe environment in which they are free to explore and in which they will not be ridiculed for attempting something different.

An agreement to take on a fictive mode allows for aesthetic doubling (Iser, 1978). The learner creates a character, a fictive time, a fictive space and a plot yet still remains himself. This aesthetic doubling allows learners to experience a different context yet analyse it through their own eyes. The dramatic contract makes it clear for the students that they step into fiction and they step out of fiction and reflect upon what the learning was about. It is this aesthetic doubling which permits the inquiry nature of learning through a drama activity.

2.1.2 Drama Allows for Mental Models to Be Constructed and Examined. It Is Multimodal and Multisensory

Drama in science education focuses on the creation of dramatic situations to be explored by the participants to find out how and why, shift perspectives, identify problems and perhaps solve them. These are all key factors in inquiry-based learning. Drama builds upon taking roles and relies on many voices. The DICE Consortium (2010) mentions "[Drama gives an] opportunity to probe concepts, issues and problems central to the human condition, and builds space to gain new knowledge about the world."

Multimodal learning theory (c.f. Kress, 2010; Van Leeuwen, 1999) suggests that the making of meaning can be supported by different semiotic resources in different modalities, which have different affordances for meaning making. Drama uses a range of semiotic modes such as visual images, voice, music, movement, embodiment and sensory input, with the body and the voice being the main instruments of dramatic expression. Furthermore, in drama these different modes appear simultaneously. This multitude of modalities can help students in their process of inquiry and may tap into modes not traditionally utilized in the science classroom (visual images, voice and writing). This multimodal inquiry can be helpful in making meaning out of abstract scientific concepts.

2.1.3 Drama Activities Are Sociocultural Activities and Allow for Scaffolding in Learning Science

Drama activities happen in social settings. They usually require group work and are thus often collaborative and interactive. Such activities allow for sociocultural learning according to Vygotsky's sociocultural theory (1978), which explains learning as a complex endeavour where social context and individual learning processes interact.

In his sociocultural theory of learning, Vygotsky suggests a zone of proximal development (ZPD) as a cognitive area close to the learner's (child's) world, which cannot be tapped into alone; yet with the support of a more competent peer or adult (such as the teacher), he or she can master this cognitive challenge. Vygotsky suggests that working in the range of the ZPD creates activity which is engaging and fruitful, because it challenges the learner within a potential learning area. Good teaching practice should target the ZPD of every learner. Scaffolding, a concept derived from Vygotsky's theory, is a teaching strategy which provides support just beyond what the learner knows and can manage alone, thus tapping into the ZPD and facilitating the learner's development (Wood, Bruner, & Ross, 1976). Vygotsky maintains that the learner makes use of mediating tools in learning. Drama can be such a mediating tool. The teacher is also a "master" mediator when supporting the learner in the ZPD. Different techniques or conventions in drama provide such scaffolding by allowing the learner to explore a new or unknown situation yet still remain in touch with his or her known world of experience and knowledge. Drama activities should be planned to suit the ZPD of participants in individualized ways. The open architecture of drama offers participants to take different roles, with different levels of challenge according to their competencies within a certain area of knowledge and their individual ZPD. In an improvisation, the different solutions to the given task can be considered as different outcomes of the individual ZPDs. Linking with inquiry-based learning, such improvisations allow the learners to construct alternative explanations and to test them. Such settings also require that the learner communicates ideas and models developed in the process of inquiry to others.

2.1.4 Drama Activities Introduce Imagination and Creativity to the Science Classroom

Vygotsky underlines the importance of imagination and creativity in children's lives. He states that imagination works with experiences and things that exist. However, the result of imagination is a creation of something that has not existed before it comes into being through the child's creative process. Play creates a zone of proximal development for the child. "In play a child always behaves beyond his average age, above his daily behavior; in play it is as though he were a head taller than himself" (Vygotsky, 1978, p. 102). Drama can support the development of creativity and the capacity of imagination. It is worth noting that both creativity and imagination are important aspects not only of art but also of science itself: it takes imagination to build a mental image of, e.g. small particles, and it takes creativity to come up with new theories and experiments to find them. In effect, imagination and creativity are key components of real scientific inquiry and could be in inquiry-based science learning.

2.1.5 Drama Activities Allow for Both Narrative and Logical-Scientific Thinking

Drama activities allow for a combination of logical-scientific thinking and narrative thinking (cf. Bruner, 1996). On the one hand, students must relate to the scientific content when designing their performance: What topics will be discussed? What scientific model will be shown? Are the topics being presented correctly? On the other hand, when students work in drama mode, narrative thinking is predominant because a story is told through the drama form. In the dialogical process of producing the drama piece, pupils flip-flop between logical-scientific thinking and reflection on one hand and the more artistic narrative thinking on the other.

2.1.6 Drama Activities Allow For Exploring Situations and Include Performance for the Classmates

It is important to note that drama activities and theatre function along a continuum with process at one end and performance at the other (DICE, 2010, p. 18–19). Process allows for exploring, sharing, crafting, presenting and assessing. Performance is part of the comparison of the results of the exploration. Classroom drama activities usually stress the process, where most learning is expected to take place. In professional theatre plays, like in science theatre, however, the learners do not participate in the creation process and are only exposed to the final product. In such cases, learning happens while watching the performance. When science theatre is performed in school context, there is usually a didactical package for the teacher and the students to elaborate, before and after the performance. In the following section, we will lift up to the front two projects where science and art are combined.

2.2 Examples of Teacher Training

As part of the S-TEAM project (a pan-European project aimed at introducing inquiry-based teaching methods into teacher education), we have created and enacted two workshops for science teachers on the use of drama tools as inquiry-based methods in science education. In the workshops, teachers were asked to participate in the drama activities as if they were the students. This is a good way for the teachers to experience first-hand how drama activities operate. We further suggest that mere participation in a drama workshop can improve teachers' presentational skills (indeed in some countries like Australia, Taiwan and Iceland, drama is integrated in the key learning area the arts). However, teachers are sometimes reluctant to let go of inhibitions and play the role of the student. We therefore suggest that throughout the workshop and before each exercise, the instructor should point out the theoretical and practical aims of the exercise. This gives the exercise a concrete aim and provides the teachers with a reason to participate in the drama activity. The general structure and content of the two workshops are presented here. Further information may be found on http://www.storyline-scotland.com/2011/03/report-from-trondheim-university-ntnu/.

2.2.1 Example 1: A Threefold Learning Loop Through a Storyline Within Teacher Education

The Scottish storyline method[1] is characterized by a fictive frame, which is based on involvement from the students' side. The storyline narrative is guided by key questions, which are open and demand exploration (such as "What does the building of a water reservoir mean for a local community?"). The key questions are formed in accordance with curricular aims. With older pupils and adult students, science loops are introduced. These are short lectures about the issue explored. Learning from a storyline is inquiry based and explorative and requires students' active participation. Table 1 introduces a planning tool for storyline projects. (More information about the water storyline projects is available on http://www.storyline-scotland.com/2011/03/report-from-trondheim-university-ntnu/)

Central to the learning process is a meta-discussion about what the group has achieved as well as a process-oriented formative evaluation. The final task in a sto-

---

[1] http://www.storyline-scotland.com/news.html

Table 1 Storyline planning tool

Key question	What the teacher does	What the students do	How to organize	Products	Aim/learning outcomes	Assessment
Question 1						
Question 2						

ryline project is a presentation or a performance where the students share the knowledge and insights established during the project.

In the S-TEAM project, we have carried out storyline projects in teacher education about issues connected with water and issues connected with exploring the universe. A typical work process in teacher education contains three learning loops. First, prospective teachers are presented with a storyline project lasting 3–4 h. At this point the aim is twofold: to experience storyline and to learn about the didactics of a storyline. The second learning loop is that the prospective teachers plan their own storyline (inspired by the one they were shown) to be carried out as a 1-day project in a secondary school. The third learning loop is to actually carry out a project and guide students in secondary school through a storyline lasting 4–6 h. The prospective students who participated in this process gave a positive evaluation of a storyline as an engaging and motivating method, where different disciplines can be involved in exploring scientific issues. They consider the third learning loop of utmost value, actually carrying out a storyline project in practice with real students in secondary school.

In the autumn of 2011, we created a storyline called "SPACE ME" about the universe, for teacher students, who plan and further develop this storyline and carry it out in their practice period. A few days before the storyline at the teacher education institution, the students were asked to send in a one-page-long text called "The universe and I." Through this piece of personal writing, the teacher students reflect on their prior experience of themes connected to the universe. On the actual day of the first learning loop, a hook is planted in the form of a teacher in role acting as a cyborg (half-human, half-robot), whom the students meet outside the learning space. The cyborg is inscribed with a text: "Press the button. Please touch!" When someone presses the button, the cyborg starts moving for a short while. When the teacher students enter the auditorium, they are met by a teacher in role as a scientific researcher, and they fill out a questionnaire on their knowledge about the universe. After the "hook" the students meet a science loop by a multimodal lecture of one teacher educator about our solar system, astronomy and space travel (such as the Voyager spacecraft from 1977). After that, the teacher students are framed as the general assembly at the United Nations and asked to create expert committees with five persons in each. The task of each committee is to design a cyborg to undertake a journey to outer space; this cyborg must be prepared to represent humanity in case of an encounter with extraterrestrial beings. Every member must have special knowledge of some field of science. They fill out intergalactic passports with a made-up name, which describes the skills and competencies of the expert. They should especially focus on the skills needed if the cyborg should encounter alien intelligence in space. The final task is to present the cyborg they have created and the skills of the expert team, which will support the cyborg on its planned journey. The expert teams present a manual for the cyborg's galactic journey (with a map of the planets, with medical advice, with physical training advice and so forth). Finally, there is a meta-reflection about the learning and an evaluation of the project. This huge theme can then be developed further within science classes (Østern, 2013; Østern, & Kristoffersen, 2015; Østern, & Østern, 2016; Østern, & Strømme, 2014).

2.2.2 Example 2: Description of a 1 or 2-Day Workshop on Drama in Science Education

This workshop has been conducted with science teachers of various disciplines and seniorities and with postgraduate students (some of whom are also teacher educators), with a typical group size of around 20 participants. The workshop commences with a short discussion of differences and similarities between the arts and the sciences. Participants write down their opinions on a paper and then share them with the class. Teachers (as do most people) tend to see the two as very different and dichotomous. The discussion is wrapped up by presenting Ashkenazi's (2006) argument that in fact science and art share similar cognitive demands and abilities of abstraction. This is the starting point and rationale of the workshop. A short lecture is then given to present the workshop's aims and the general background of drama as an inquiry-based learning method.

At this point participants are asked to stand up and push chairs and tables back to create a working space. A short warm-up, icebreaker exercise is conducted followed by the explanation and the enactment of a fictive contract. Teachers are then asked to perform several of the drama conventions mentioned earlier in groups. During these exercises the elements of inquiry-based learning in drama activities are stressed. The teacher's role is also discussed with examples given from incidents that occur in the activities.

Finally, teachers are asked to write and enact their own short lesson plan that includes drama. These must include a rationale and difficulties they anticipate in class (too much noise/no cooperation). These exercises can bring out teachers' fears and scepticism and allow the instructor to address them.

Following written and oral feedback and participants' reflections, it seems that the pre-service teachers highly appreciated the new techniques and ideas to which they were exposed, as is seen in the following reflection:

> I choose the [drama] workshop as one of the most influential parts of the training. I feel that the workshop gave me tools to communicate with my future students in a special way ... I feel that activities from the workshop correspond with students in a different way and allow even the quietest of students to join in. I also felt that during the workshop, I, myself, could act as a student – I felt at liberty to express myself and my feelings. I think that I can create a better initial contact with my students and show them the lighter side of physics and robotics.

A pre-service teacher educator testified that in the 2 years she taught tools from the workshop, the students "claimed it was the most memorable and enjoyable part of their teacher training and that they will use these exercises in class." From the workshop discourse, we found that the pre-service teachers also learnt specific science content by performing the exercises. In one of the activities, for example, the participants were asked to form two circles representing the cell wall and membrane of a plant cell. It was only then that one of the students realized that she did not know what was in the space between the two, despite having learnt the topic in biology classes. The workshop not only allowed the students to learn teaching tools but also allowed them to sharpen their understanding of the content knowledge of their discipline.

While the pre-service teachers highly appreciated the workshop, it remains to be seen if they would actually use the tools provided. Lack of experience might inhibit them from using such tools, and this issue could be addressed. However, some of the teachers successfully implemented the activities: "We thought the students would be reluctant…but [when we tried,] everyone cooperated and the topic became clear and interesting."

From the workshop discourse, it seems that the in-service high school teachers were more reserved about adopting tools from the workshops. Many felt that they cannot use the tools due to the pressure of preparing students for their matriculation exams. However, discussion revealed that many have and do in fact use drama activities in class. One typical feature of the workshops with in-service teachers was that after each activity, a lively discussion often developed, which focused on content, accuracy, implications and suggestions for improvement. Since these teachers are highly experienced, they had a lot to say about the learning and could better imagine these exercises in the classroom context than their pre-service counterparts. Such discussions serve as a good demonstration for the arousal and engagement drama activities can cause. They also demonstrate how drama activities can be used to deepen and sharpen embodied knowledge.

3   Illustrations of Drama in Science Education

In this section we will investigate practical options of introducing drama to the science classroom, which foster an inquiry-based learning environment. We will suggest exercises and examples that can be used in class.

3.1 The Dramatic Contract

As mentioned earlier, when using drama in science education, there needs to be a clearly defined and understood contract about the agreement to pretend. The contract clarifies the boundaries, lays down clear rules and gives a control mechanism, with the aim of creating a secure yet focused environment which allows learners to test, investigate and dare. O'Toole and Dunn (2002) have listed some suggestions about the content of such a dramatic contract. They underline that this contract does not necessarily need to be written, but it must exist. The main aspects of the dramatic contract are:

- In the dramatic context, we must all work together to make the fiction work, that is, taking a role, staying in the role and accepting the make-belief. This does mean that students cannot explore within their role. On the contrary, students should be able to try things out (movement, emotions, relationships) but within certain agreed limits (e.g. if a student espouses the role of a neurotransmitter, she should try out different actions within agreed boundaries. If she sees the boundaries are unsuitable, in the next run of the drama, she can ask to change these rules).
- We should use our imagination to transfer to other places and times.
- We understand that the roles taken and the drama stories explored have a learning purpose in the curriculum.
- We are having fun – seriously. We must take care of both each other and the drama.
- At a particular signal from the teacher, everybody will immediately "freeze."
- There are three stages to the activity: we enter the drama, we exit the drama and we reflect on the drama.
- The teacher can take roles within the drama.

3.2 Process Drama

After establishing a dramatic contract, the drama activity can take a great number of forms (or genres). Process drama, one of the main forms of educational drama, stresses the process (vs. the performance) and is suitable for both simulations and analogies. The genre is characterized by a starting point or a pretext from which the inquiry process begins. This can be introduced in the form of a question, a challenge, a text from a newspaper, a story, etc. Process drama is improvised and does not require a pre-written manuscript. The progress of the exploration in process drama has a dramaturgy of episodes: through the creation of dramatic scenes, students explore the questions posed. Dilemmas, choices, decisions, parallels (long

ago – now – in the future) might be juxtaposed in process drama through mono-
logues where we meet the same character in different phases of life. The conventions can also be used as small dramatic moments in the classroom without the
development of a whole process drama.

An extensive toolkit of conventions (or techniques) is available for process drama
(Neelands, 2000). Four conventions are presented here:

3.2.1 Still Images (Statues, Tableaus)

Working in small groups of 3–5, pupils are asked to form three still images with
their bodies as materials. The still images sum up the main characteristics of a phenomenon, a story, an event or an attitude. The image can be realistic or abstract. It
is useful to stress that images with dramatic tension are more interesting, that the
images should mirror the pupils' interpretation and that it should be possible to
maintain the positions chosen for enough time for the rest of the pupils observe and
comment on the meaning of the images. A short time to plan the images (about
5 min) is given, the images are presented and finally the rest of the group responds
and reflects. This convention can be used for analogies. Students can, for example,
be asked to show their version of the change of the state of matter (e.g. boiling). The
other students and the teacher can reflect on the strengths and weaknesses of the
model presented in the exercises. The convention can also serve as a simulation. In
the storyline project (described below), students were asked to form family groups
and form a still image of one moment where a threat was met from water (mud-
slides, tsunami, etc.). The exercise together with the subsequent discussion allows
the teacher to understand how and to what extent the content has been assimilated.

3.2.2 Hot Seating

In hot seating one pupil volunteers to take on a role of a character or an entity whose
attitudes or nature we are exploring. The rest of the group can pose questions to the
pupil in role, who answers according to what the character might think about this
issue. If used to explore a natural phenomenon, for example, questioning a predator
or a prey in ecology, stress must be given in the debriefing to warn of
anthropomorphism.

3.2.3 Teacher in Role

This is one of the main conventions in process drama. The teacher enters into a role
presenting a figure with certain attitudes and values and works in a fictional context
with the pupils as participants. In our storyline project about water (see below), one
teacher in role described a quick clay landslide as the grandchild of a survivor;
another teacher in role was from an international humanitarian organization

inspecting the floods in Pakistan from a helicopter; a third teacher in role placed herself as a wife and mother in Maldives when the tsunami hit in 2004.

3.2.4 Conscience Alley

This is a whole group activity. Pupils stand in two parallel rows and are asked to take a standpoint on a certain topic and form a sentence reflecting this standpoint, like a thought going through one's head. When one pupil (either in role or as herself/himself) walks through the alley, the other pupils repeat the sentence they chose. The person walking through the alley listens to the voices and at the end of the alley says what decision s/he will make. This can be repeated several times with different walkers.

3.2.5 Choices and Dilemmas

Another variation on choices and dilemmas: on a fictive line in the classroom, pupils can place them according to their view on a certain dilemma in science. It could be, for instance, if they are afraid of climate change. Then they argue (not in role) for the place they have chosen.

3.2.6 Monologue in Role

A fictive or historical scientist can be presented via a monologue, based on facts found by the pupils. It could be a story told by Marie Curie.

3.3 Drama Can Promote Depth of Learning and Holistic Understanding

The digital revolution has changed the science classrooms, and the digital media offer use of avatars (figures), games and three-dimensional models that demand skill to stage fictive scenes in order to explore the not yet known.

We have mentioned earlier that drama offers embodied learning (Wilson, 2002). The point is that the body, mind, cognition and emotion are engaged in inquiry-based learning processes. A special aspect is the need to clearly state ethical aspects of science by asking "Who cares?" and "So what?" In a Norwegian white paper from 2015 (NOU, 2015), an expert panel suggests what school in Norway faces of challenges in a future perspective. The expert panel mentions three major challenges: climate changes, multiculturalism and that young people do not master their lives. All these themes are major themes in science and can be explored in depth by means of drama. The expert group asks for more practice and more aesthetic activities in school and more inquiry-based projects that might enhance depth of learning. Depth of learning means understanding of totalities, not merely to be able to retell the curriculum stuff.

4   Teacher Training for Drama in the Science Classroom

In this section we will discuss aspects of training teachers to enact drama in the science classroom as an inquiry-based activity. We provide an overview of the roles and responsibilities of the teacher in such an environment.

Performing an inquiry-based drama activity in the classroom is different from teaching a traditional lesson. Such activities are more open ended and give students more autonomy. The teacher must be aware of his role as a mediator rather than an absolute source of knowledge, a role that allows students to perform their inquiry. The teacher must also introduce the pupils to drama skills they might not be acquainted with. The following factors should be taken into consideration by teachers in order to run a successful drama activity. These factors should also be stressed in teacher training and teacher professional development when introducing drama activities.

4.1   Before the Activity

Before a drama activity begins, the teacher must brief the students as to what is going to happen. While some drama activities are open ended and students are given free choice, the boundaries, limits and objectives in which the inquiry activity operates, as well as its goals and its assessment, should be clear to the teacher and made clear to the students. Part of the briefing is, of course, the dramatic contract described earlier.

4.2   During the Activity

The teacher has an important role in supervising the activity. During a drama activity, many questions may arise. These may be about material already learnt or concepts new to the students. The teacher should be attentive and add information during the activity as needed or remember questions which might be best answered in front of the whole class after the end of the activity. In one of the activities we conducted, for instance, students created a dramatic model of a plant cell. Only during this inquiry did the question arise "what is there between the cell wall and the membrane?" This was despite a picture of the cell being shown on the board many times. Such moments can serve to reveal the class' understanding of a topic.

4.3   After the Activity

At the end of the activity, debriefing should take place. The debriefing starts with de-rolling, taking the group out of the fiction. The debriefing allows for a repetition and a summary of the activities seen, a discussion about the validity and correctness of the models presented and linking of the activities to the learned material in class. This metacognitive reflection is important for the students' learning, to articulate the experience and the contribution to new knowledge, change of attitude and agency.

4.4   Skills Needed from the Teacher

The teacher who wishes to introduce drama to the science classroom needs to have a positive attitude towards dramatic elements, mainly regarding the value of the forms of knowledge that might be strengthened through this aesthetic approach. It is also necessary for the teacher to have some understanding of the importance of quality in the dramatic expression, mainly in a strong focus on the task – also in the playful mode.

Despite the many advantages of introducing drama activities to the science classroom, science teachers may be reluctant to do so due to lack of skill (Alrutz, 2004). Ideally, science teachers should take some courses in drama during their teacher education or later in continuing professional development. Alternatively, we suggest that teachers with no drama experience should have support from a mentor or colleague with drama competencies. However, many of the techniques listed above can be applied using general teacher thinking, and many science teachers introduce drama naturally in a step-by-step manner (Dorion, 2009). The teacher needs to structure the task clearly, divide the class into smaller groups, set a time limit and describe how the task will be presented.

Introducing drama may differ between different age groups. Children at primary level can connect to their own dramatic play, while older children might need good models for role taking. Drama could be introduced in a progression. In the beginning "everyone should do everything at the same time" (with no solos for beginners!). Then work can be done in small groups of 3–5 pupils. Finally, solos may be introduced. We believe that when done properly, drama in science education is suitable across all age ranges, and indeed examples can be found at primary (e.g. Mcgregor, 2012; Warner & Andersen, 2004), junior high (e.g. Abrahams & Braund, 2012; Østern & Peleg, 2014) and secondary levels (e.g. Abrahams & Braund, 2012; Peleg et al., 2015).

Drama activities do not necessarily fit all topics taught, nor are they necessarily suitable for a specific point in time. The activities may be used as an introduction to a topic, a conclusion to a topic, as an inquiry activity to deepen the topic learned, etc. Teachers should use their discretion to decide when to best utilize a drama activity to fit their general goals.

Another concern brought up by many teachers is the potential for disorder and too much talking in the classroom. As drama activities are collaborative and dialogical, talking is necessary for exploration and for the formation process. The dramatic contract gives the teacher the necessary boundaries for control. Also, the "freeze" sign may be used as a tool to control the situation at any time. The engagement of the students may be identified through focused work and lots of talking. Nevertheless, it will take some time for students to get used to the introduction of drama into the classroom. We recommend that drama is introduced in small portions. The still image convention, for instance, has a low make-believe threshold and is a good starting point. The culture of drama and the dramatic contract are slowly constructed, thus building the confidence of the students.

4.5   Focus on the Pupil's Learning and on Formative Assessment

One of the main goals of assessment in teaching is to make students' learning (or lack thereof) become visible. From our analysis of the characteristics of drama as a learning medium above, we consider one of the contributions of drama to be that it enhances learning through embodied cognition. To this extent, traditional summative assessment based on the curriculum is one way to assess the success of a drama activity and to control the knowledge acquired. Yet, while traditional testing can provide insight into whether the activity was successful in teaching specific topics, it cannot provide a full picture of the learning outcomes, especially the affective and attitudinal ones. We suggest that more creative assessment tools are utilized, such as writing in role where students write in the role of a certain character.

Formative assessment should also be encouraged in drama activities. The paradigmatic shift from focus on teaching to focus on learning is one consequence of the sociocultural and socio-constructivist view of knowledge development. This focus also stresses learning as an ongoing process in which formative assessment can provide good scaffolding. Formative assessment is given during the work process, in order to scaffold the student's learning. In order to make use of formative assessment, the criteria for competence achieved must be formulated and made known to all. In formative assessment the learner gets "information about the gap between the actual level and the reference level of a system parameter which is used to alter the gap in some way" (Ramaprasad, 1983, p. 4). The response is the core in formative assessment. It is a diagnostic response and a system for learning. Joan Hattie and Helen Timperley (2007) write that formative response should feed up, feed back and feed forward: Where is the goal? Where am I? How shall I continue?

Three forms of formative assessment are central in developing an assessment culture in the classroom: teacher assessment, peer assessment and self-assessment. In a process drama activity, all three may be incorporated. To teach science teachers to carry out formative assessment will most likely have a profound effect on their teaching of science. The literature found regarding formative assessment is very rich and extensive. Already in 2000, Shepard (2000) summed up the main thoughts within the field assessment. Assessment in science education has been addressed, for instance, by Black (2017).

In the Norwegian framework curriculum for science (LK06 [Knowledge promotion], 2006), the notion of the budding researcher is used. This concept invites to inquiry-based learning. An example of how this is applied in science teaching is shown in the project Seeds of science/Roots of Reading (http://scienceandliteracy. org/), in Norway, carried out by Ødegaard, Haug, Mork, and Sørvik (2016). The catch words for this inquiry-based project are as follows: Do it! Talk it! Read it! Write it! The challenge for the teachers is to find multimodal variations in inquiry. The phases of inquiry are to prepare, to gather data, to discuss and to communicate. Svendsen (2017) has in her longitudinal professional development project with science teachers successfully applied IBST/L strategies in formative interventions over 3 years in Norwegian school context.

5 Conclusion

In this chapter we provided a framework for teaching and learning science through drama. We suggest that learning science through drama offers inquiry-based learning which functions through narrative and is multimodal and multisensory and has sociocultural features. We also articulated what skills are required in order to include drama as a learning medium in class and provided specific examples, which can benefit science teacher educators. Although it is in use in many cases, drama in science education is far from reaching its full potential. One of the challenges is the lack of training for teachers in drama as a learning medium in science education. We suggest that this chapter can provide the necessary background for science teacher educators wishing to include drama as a learning medium in science education.

References

Abrahams, I., & Braund, M. I. (2012). Performing science: Teaching chemistry. In *Physics and biology through drama*. London: Continuum.

Alrutz, M. (2004). Granting science a dramatic license: Exploring a 4th grade science classroom and the possibilities for integrating drama. *Teaching Artist Journal, 2*(1), 31–39.

Ashkenazi, G. (2006). Metaphors in science and art: Enhancing human awareness and perception. *Electronic Journal of Science Education, 11*(1), 3–9.

Black, P. (2017). Assessment in science education. In K. Taber & B. Akpan (Eds.), *Science education an international course companion* (pp. 295–309). Rotterdam, The Netherlands: Sense Publishers.

Bruner, J. (1996). *The culture of education*. Cambridge, MA: Harvard University Press.

DICE Consortium. (2010). *Making a world of difference. A DICE resource for practitioners on educational theatre and drama*. Belgrade, Serbia: DICE Consortium.

Dorion, K. R. (2009). Science through Drama: A multiple case exploration of the characteristics of drama activities used in secondary science lessons. *International Journal of Science Education, 31*(16), 2247–2270.

Hattie, J., & Timperley, H. (2007). The power of feedback. *Review of Educational Research, 77*, 81.

Iser, W. (1978). *The act of reading. A theory of aesthetic response*. Baltimore/London: The John Hopkins University Press.

Kress, G. (2010). *Multimodality: A social semiotic approach to contemporary communication*. London: Routledge.

LK06. (2006). *Natural science subject curriculum*. Oslo, Norway: Utdanningsdirektoratet. Retrieved September, 2011, from http://www.utdanningsdirektoratet.no/upload/larerplaner/Fastsatte_lareplaner_for_Kunnskapsloeftet/english/Natural_science_subject_curriculum.rtf

McGregor, D. (2012). Dramatising science learning: Findings from a pilot study to re-invigorate elementary science pedagogy for five-to seven-year olds. *International Journal of Science Education, 34*(8), 1145–1165.

McSharry, G., & Jones, S. (2000). Role-play in science teaching and learning. *School Science Review, 82*(298), 73–82.

Neelands, J. (2000). In T. Goode (Ed.), *Structuring drama work*. Cambridge, UK: Cambridge University Press. (First published 1990).

NOU 2015-8. (2015). *Fremtidens skole – fornyelse av fag og kompetanser* [The school of the future – Renewal of subjects and competences]. https://www.regeringen.no/no/dokumenter/nou-2015-8/id2417001/sec1

O'Toole, J., & Dunn, J. (2002). *Pretending to learn: Helping children learn through drama*. Port Melbourne, Australia: Pearson Education Australia.

O'Toole, J., & Dunn, J. (2013). *Pretending to learn- teaching drama in the primary and middle years*. First published in 2002. French Forest: Pearson Education Australia. The third revised edition is a Kindle digital edition.

Ødegaard, M., Haug, B. S., Mork, S. M., & Sørvik, G. O. (2016). *På forskerføtter i naturfag* [On researcher feet in science education]. Oslo, Norway: Universitetsforlaget.

Osborne, J., & Dillon, J. (2008). *Science education in Europe: Critical reflections* (Vol. 13). London: The Nuffield Foundation.

Osborne, J., Simon, S., & Collins, S. (2003). Attitudes towards science: A review of the literature and its implications. *International Journal of Science Education, 25*(9), 1049–1079.

Østern, A. L. (2013). Close-up of the craft of making a multimodal lecture for teenagers about the universe: The importance of a meaningful context and supporting colleagues. *Reflective Practice, 14*(6), 801–813.

Østern, A.-L. (2014). *Dramaturgi i didaktisk kontekst*. [Dramaturgy in educational context] Bergen, Norway: Fagbokforlaget.

Østern, A. L., & Kristoffersen, A. M. (2015). Combining art and science in exploration of humanity and the universe: Teenagers' storied experience of the universe played back in improvisational theatre in a learning context. *Applied Theatre Research, 3*(3), 251–270.

Østern, A.-L., & Peleg, R. (2014). Kroppslig læring om gravitasjon. Dans, mime og naturfagpå ungdomsskolen [Bodily learning about gravity – Dance, mime and science in junior high school]. In T. P. Østern, & A. Strømme (Eds.), *Sanselig didaktisk design SPACE ME* [Aesthetic pedagogical design SPACE ME] (pp. 129–152). Bergen, Norway: Fagbokforlaget.

Østern, T. P., & Østern, A.-L. (2016). Storyline as a key to meaningful learning arts and science combined in SPACE ME. In P. Mitchell & J. M. McNaughton (Eds.), *Storyline: A creative approach to meaningful learning and teaching* (pp. 116–135). Cambridge, UK: Cambridge Scholars Publishing.

Østern, T. P., & Strømme, A. (Eds.). (2014). *Sanselig didaktisk design SPACE* ME [Aesthetic educational design]. Bergen, Norway: Fagbokforlaget.

Peleg, R., Katchevich, D., Yayon, M., Mamlok-Naaman, R., Dittmar, J., & Eilks, I. (2015). The magic sand mystery. *Science in School, 32,* 37–40.

Peleg, R., Yayon, M., Katchevich, D., Mamlok-Naaman, R., Fortus, D., Eilks, I., et al. (2017). Teachers' views on implementing storytelling as a way to motivate inquiry learning in high-school chemistry teaching. *Chemical Education Research and Practice, 18,* 304–309.

Ramaprasad, A. (1983). On the definition of feedback. *Behavioral Science, 28,* 4–13.

Selander, S. (2017). *Didaktiken efter Vygotskij* [The didactics post Vygotskij]. Stockholm, Sweden: Studentlitteratur.

Shepard, L. A. (2000). The role of assessment in a learning culture. *Educational Researcher, 29*(7), 4–14.

Svendsen, B. (2017). *Inquiries into teacher professional development. A longitudinal school-based intervention study of teacher professional development (TPD) conducted in the frame of a cultural historical activity theory perspective (CHAT).* Dissertation, Trondheim, Norway: Norwegian University of Science and Technology.

Van Leeuwen, T. (1999). *Speech, music, sound.* London: Palgrave Macmillan.

Vygotsky, L. S. (1978). *Mind in society.* Cambridge, MA: Harvard University Press.

Warner, C. D., & Andersen, C. (2004). "Snails are Science": Creating context for science inquiry and writing through process drama. Youth Theatre Journal, 18(1), 68–86

Wilson, M. (2002). Six views of embodied cognition. *Psychonomic Bulletin & Review, 9*(4), 625–636.

Wood, D., Bruner, J. S., & Ross, G. (1976). The role of tutoring in problem solving. *Journal of Psychology and Psychiatry, 17,* 74–76.

Part II
Familiarizing Teachers with Motivational Approaches and Scientific Literacy Goals for Inquiry Based Learning

Using Motivational Theory to Enrich IBSE Teaching Practices

Hanne Møller Andersen and Lars Brian Krogh

1 The Relation Between IBSE and Motivation

The need for a more motivational school science that stimulates students' interest in science and science careers has been iterated in many reports (Osborne & Dillon, 2008; Osborne, Simon, & Tytler, 2009; Rocard et al., 2007). These reports differ in the nature of their concern, e.g. whether they are concerned about recruitment to science careers or worry about the many students that leave school science more or less alienated from science and science-related issues in everyday life. However, both educo-political reports and research reviews agree that science teachers' instructional practices have crucial influence and that professional development is needed to meet the challenge (Osborne & Dillon, 2008; Rocard et al., 2007). Osborne and Dillon explicitly draw attention to "ways of organizing the teaching of science that address the issue of low student motivation" (Osborne & Dillon, 2008, p. 8). They recommend IBSE approaches as solution to the problem of low student motivation:

> Emphasis on engaging students with science and science phenomena is best achieved through opportunities for extended investigative work and 'hands-on' experimentation and not through a stress on the acquisition of canonical concepts. (Osborne & Dillon, 2008, p. 9)

Independent reviews on laboratory work and student motivation make more moderate and differentiated claims concerning the motivational benefits of IBSE-oriented approaches. In a literature review on laboratory work, Singer et al. (Singer, Hilton, & Schweingruber, 2006) found "some evidence" for attainment of

H. M. Andersen (✉)
Aalborg Katedralskole, Aalborg, Denmark
e-mail: hma@aalkat-gym.dk; HMA@Katedralskolen.dk

L. B. Krogh
Teacher Department, VIA University College, Aarhus, Denmark
e-mail: labk@via.dk

© Springer International Publishing AG, part of Springer Nature 2018
O. E. Tsivitanidou et al. (eds.), *Professional Development for Inquiry-Based Science Teaching and Learning*, Contributions from Science Education Research 5, https://doi.org/10.1007/978-3-319-91406-0_5

interest in science by "laboratory experiences" but stressed that some instructional IBSE approaches seem better for this purpose than others. Similarly, a recent review of studies on students' interest in science concludes:

> ...a rather strong impression that emerges from this review is that most 'inquiry-based' or 'problem-based' interventions have positive effects on students' Interest/Motivation/ Attitudes, while most 'hands-on' activities which do not require as much reflection, do not. (Potvin & Hasni Potvin & Hasni, 2014, p. 103)

In short, research evidence seems to establish the motivational potential of reflective IBSE approaches. To unfold this motivational potential, teachers must have capacities to design and enact a range of IBSE approaches. But more specifically, they should be able to include motivational strategies in their planning of IBSE, be aware of individual students' motivational states, and be capable of responding to motivational issues that arise in classroom situations. Such motivational knowledge and strategies are not usually addressed explicitly in Danish teacher training programs.

The problem with students' motivation may be targeted in several ways, but our basic idea has been to facilitate reflection and enhance motivational action by introducing teachers to elements of contemporary motivational theories. Here we concur with the view that "science educators have much to learn from the growing body of literature on the study of motivation" (Osborne, Simon, & Collins, 2003, p. 1073). However, we share Shulman's concerns about the usefulness and applicability of theoretical, propositional knowledge for teaching practice: "Although principles are powerful, they are not particularly memorable, rendering them a problem to apply in particular circumstances" (Shulman, 1986, p. 11). Acknowledging the tension between theoretical and practical knowledge, the real challenge for in-service teacher training is not to engage teachers with motivational theory but to help them integrate and transform such knowledge for use in science classrooms. It takes a consistent focus on transforming theoretical knowledge in order to promote a "propositionally interpreted practical knowledge" in schools (Thiessen, 2000).

In this chapter we will introduce our "pragmatic use" of motivational theory, which directs us to extract motivational constructs relevant for science education and provides motivational foci for teachers' planning of IBSE practices. Using data from a recent R&D project with science teachers from Danish upper secondary school, we will describe how much needed interactions between theoretical and practical knowledge can be designed for in a teacher development program. The theory-practice-transformation was facilitated by deliberate use of strategies like reflective writing, research-based teacher motivational awareness activities in participants' own classrooms, videos from teachers' classrooms, and structured video clubs (Sherin, 2007). We describe how these course elements were applied and indicate how the teachers engaged with them and benefitted from them. As such, our chapter attempts to embrace the potential enrichment of IBSE through motivational theory, outline elements of an in-service teacher development program, and indicate what can realistically be achieved within the context of a shorter (five workshops) professional development program "Enhancing Teachers' Capacities to Motivate".

The program was derived from two cycles of design-based research, but in the present chapter, we confine ourselves to data from the first cycle, where eight science teachers from different schools participated.

2 A Pragmatic Approach to Motivational Theory for Teaching

Two major considerations have been subsumed here: (1) what motivational theories/content should be introduced to teachers, and (2) in what form should this knowledge be made available to teachers?

Motivation is multifaceted, and motivational theories are plentiful (Ford, 1992). To illustrate this point, Ford, in his 1992 review, found 32 distinct motivational theories. Almost exclusively, they derive from social psychology, and typically they are built around a single/a few constructs that can be controlled for and measured within laboratory conditions. We will therefore have to question the applicability of experimentally based motivational theories and select theories and constructs as being suited for IBSE and science classrooms.

A pragmatic selection of motivational knowledge should emphasize its usefulness for teachers. Reasonable criteria would be: Is it a viable set of theories? Is it sufficient to embrace the complexity of students' motivation in science classrooms? Is it plausible that teachers will grasp it? Does it facilitate teachers' attention, planning, and actions towards students' motivation? These criteria have restricted our literature scrutiny to sources communicating motivational knowledge for use in education. The most relevant sources have been general textbooks on motivation in education (e.g. Brophy, 2004; Pintrich & Schunk, 2002), psychologists extracting educational implications of motivational theories (Ames, 1992; Pintrich, 2003), reports of motivational intervention projects in science (e.g. Cherubini, Zambelli, & Boscolo, 2002; Martin, 2008; Stipek, Givvin, Salmon, & Macgyvers, 1998), and content descriptions of previous pre-service science teacher training programs (e.g. Anderman & Leake, 2005). The literature review leads us to conclude:

- Several motivational theories are necessary to capture educational complexity.
- A sufficient set of theories normally includes self-efficacy, self-determination theory, attribution theory, goal orientation theory, and, very often, expectancy-value theory and interest theories.
- Use of a set of complete motivational theories is unrealistically demanding. A set of selected motivational elements and constructs may do a better job. This view will be elaborated below.

Anderman and Leake describe the proliferation of motivational theories and constructs that have been fruitful to researchers but "leaves the field dense and impenetrable to others" (Anderman & Leake, 2005, p. 192). To make these theories on motivation more accessible to teachers and educators, they suggest that the content should be completely reorganized:

Rather than teaching students about theory per se, our approach focuses on principles of motivation, and we group constructs and ideas from various theories, organizing them within a larger framework. (Anderman & Leake, 2005, p. 192)

The framework of Anderman and Leake (2005) is built around a highly restricted set of three motivational constructs (autonomy, belonging, competence), and these constructs are broadly conceptualized. Elements and implications of other theories are pragmatically included. We do agree that a motivational construct-driven approach is viable, potentially sufficient, and more functional than presenting teachers to complete relevant motivational theories. We also concur with the idea of subsuming closely related constructs – with almost identical implications for practice – into one pragmatic construct, e.g. merging self-efficacy (mastery expectations) and competence (perceived mastery in situ). However, we find that Anderman and Leake's framework goes too far in their effort to reduce complexity, e.g. leaving out vital aspects of task value and interest theories. Table 1 presents our balanced pragmatic choice of motivational constructs and related recommendations for teachers' motivational foci and practice in relation to IBSE (and other activities in science classrooms). All recommendations are general in nature, leaving recontextualization to specific science classrooms; to inquiry activities; to certain groups of students, etc.; to teachers and teacher trainers. This is a critical aspect that must be taken into account in the design of development programs and in-service teacher training.

3 Motivational Constructs and Recommendations for Practice

Our overview of pragmatic motivational constructs and related foci for teaching practice is conceptualized as the CARTAGO framework in Table 1. Looking at the motivational mechanisms, it is clear that two different time scales are operating: constructs like autonomy and task value may be addressed and related student motivation installed within a single situation/task, while self-efficacy, relatedness, causal attributions, and goal orientations can only be built through consistent efforts over longer periods. Looking at the recommendations for practice, it should be noted that the constructs have implications on various levels of science teaching: some aspects implicate the whole *learning environment* (e.g. aspects of relatedness), some relate to *activities and tasks* (aspects of task value), and others are *interactional* by nature (e.g. feedback). In the literature, a richness of recommendations for practice has been identified for each motivational theory and its core construct(s). Recommendations for some constructs tend to overlap, e.g. competence and goal orientation. Generally, the constructs tend to enhance each other instead of undermine each other. However, one exception is that offering extreme autonomy to students may threaten lower-achieving students' sense of competence.

The motivational CARTAGO framework is relevant for science teaching in general, and it is a valuable tool for motivational planning and analysis of IBSE

Table 1 Motivational framework based on selected motivational constructs (CARTAGO)

CARTAGO framework	Major recommendations for a motivational classroom practice[a]
Competence/self-efficacy	Match tasks to students, so everybody regularly perceives themselves as competent
	Instigate proximate goal setting (moderately challenging but achievable goals)
	Modelling of targeted competences (peers or teacher)
	Provide formative feedback
	Apply varied assessment methods
Autonomy	Provide meaningful choices in tasks, assignments, students' collaboration
	Involve students in decision-making
	Stimulate students' self-regulation and teach it explicitly
	Minimize teacher control
Relatedness	Recognize students as individuals
	Foster an inclusive learning environment with mutual respect, recognition of diverse contributions, etc.
	Arrange for student interaction and interdependence in tasks
Task value	Plan tasks and themes so they relate to students' lifeworlds and allow students to perceive personal relevance
	Enhance students' curiosity
	Enter active and dynamical elements into tasks and assignments
	Plan for novelty and variation
Attributions to success or failure	Emphasize that success/failure in all major aspects depends on student-controllable factors (e.g. effort, tools that can be learned, etc.)
	Explicate that progress in learning is incremental, and knowledge tends to be domain-specific
	Give achievement feedback to students in private (particularly surprisingly negative feedback)
Goal Orientations mastery vs. performance	Let students establish self-referenced goals in science
	Avoid competition and discourage social comparison (e.g. avoiding unnecessary grading)
	Provide feedback on personal progress (task oriented and criterion referenced)
	Be flexible with time – students may need different amounts of time to master the stuff
	Establish that making mistakes is natural and that it has great learning potential

[a]The recommendations are drawn from the literature on motivational knowledge for use in education (e.g. Ames, 1992; Anderman & Leake, 2005; Bandura, 1997; Palmer, 2009; Pintrich, 2003; Reeve, 2002; Swarat, 2008)

approaches. Just to indicate this potential, we can use the framework to characterize the motivational affordances of a "traditional" subject-centred cookbook inquiry activity, where the aim is to verify some scientific law or principle already known by the students. The cookbook approach leaves little autonomy to the students, and the subject-centeredness reduces the chance of students perceiving personal relevance and interest (task value). Since all students are working on the same task, competence matching will be accidental, and students will have no chance of establishing self-referenced goals for the activity. These kinds of investigations are typically done in student pairs, so there will be some student interaction but no planning for interdependence. Since students typically focus on cookbook procedures, science teachers will have little opportunity to provide formative feedback on personal progress or conceptual understanding. This example illustrates how the CARTAGO framework can pinpoint the motivational weaknesses of traditional inquiry practice; in addition, the framework suggests alternative approaches probably leading to more motivational teaching.

4   Engaging Science Teachers with Motivational Constructs in In-Service Training

In this section we will describe some major design principles and strategies used to facilitate teachers' transformation of theoretical inputs into classroom practice. Along the way we will also give examples of teachers' responses and evaluation of various elements. These data derive from interviews, questionnaires, essays, and videos (classroom and workshops) from our development program "Enhancing Teachers' Capacities to Motivate".

4.1   Overall Design Principles of Our Professional Development Program

Our intervention design was inspired by the interconnected model (Clarke & Hollingsworth, 2002; Nielsen, 2012), which is a synthesis of contemporary models of teacher professional development. According to this model, teacher professional growth derives from complex interactions between contributions from external workshops, collaboration with teacher peers, and classroom trials and evaluation of students' outcomes. Workshops on motivational constructs and strategies may be relevant to transcend existing practices, but a professional development program must also make deliberate use of the other domains (peer collaboration, classroom trials, etc.) to facilitate negotiation, transformation, or redirection of workshop inputs for practice. In the interconnected model, two fundamental processes interrelate the domains and drive teacher growth: enacting classroom trials and reflecting on inputs, peer discussions, and classroom enactments. Accordingly, an adequate teacher professional development design must provide a richness of enactment and (co)-reflection affordances. Deliberate use of all domains and targeted enactment/reflection activities are the core principles of our design. The structure of our intervention and the emphases of workshops and activities are indicated in Fig. 1.

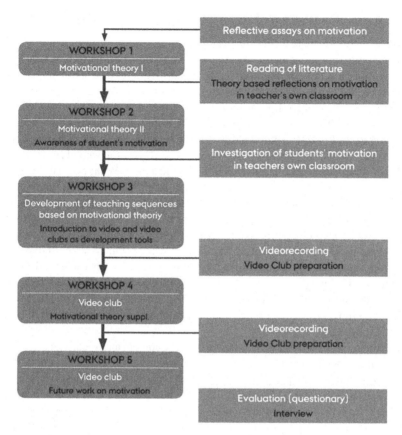

Fig. 1 Structure of teacher professional development program for motivational practice. The program duration was a semester

Motivational theory related to CARTAGO constructs was introduced at the first workshops. Workshops included multiple collaborative activities, e.g. structured discussions, co-planning of motivational teaching, and video club sessions. As clearly demonstrated in Fig. 1, these contributions consistently interacted with practice-related activities undertaken by participants at their own schools in the intervals between workshops.

4.2 Theory-to-Practice Strategies

In the next sections, we will present some of the activities and strategies that we have applied to stimulate teachers' reflections on their transformation of motivational theory into practice and reflections on students' motivation. A more detailed description of the structure and content of the development program and a description of specific activities are available on the S-TEAM website.[1]

---

[1] https://www.ntnu.no/wiki/download/attachments/8325736/Deliverable%206b%20October%20 2010.pdf?version=1&modificationDate=1297848434000&api=v2

4.3   Reflective Writing: Investigation and Motivation

Reflective writing in terms of logs/diaries, journals, and essays is a recognized tool within teacher preparation and in-service training (Burton, Quirke, Reichman, & Peyton, 2009). We have used reflective writing activities for different purposes. As a workshop pre-activity, we asked teachers to write reflective essays in order to elicit their initial ideas about student motivation and invite their voices and practical knowledge into the course. Here the writing prompts were: "Describe and reflect on one situation where you succeeded in stimulating students' motivation and another where you did not", "What do you do in your teaching to stimulate students' motivation?" and "What kind of motivational strategies do you use in your teaching and which are your favourites?" After the introduction of motivational theory, we used reflective writing as a tool to help teachers make meaning of motivational theory and take the first step towards an interpreted motivational practice. The writing was prompted with: "Give examples from your teaching that illustrate how motivational constructs and theories are enacted in your classroom". Typically, participants' shared their reflections for joint discussion at the following workshop. Sometimes we, as teacher trainers, provided overviews over participants' strategies and drew attention to typical or critical aspects.

All reflective texts written by the teachers were content analysed. The pre-activity essays on teachers' motivational strategies showed that they all had a fund of particular topics that they perceived as inherently motivational. These were mainly topics related to students' everyday life. All but one of the teachers mentioned "exciting" activities as important motivational vehicles, and most teachers pointed to the motivational effect of students doing experiments. So, prior to this intervention, the participants tended to ascribe motivational benefits to students' practical work. A few teachers attended to aspects of the learning environment as motivational, and a single teacher tried to stimulate motivation through supporting students' self-worth.

In general teachers found the reflective writing fruitful, as illustrated by a typical comment from the end evaluation: "Well, they are necessary to make you reflect and they helped you focus on important points". Not surprisingly, teachers wrote longer and more detailed essays when the reflective prompts allowed them to be more descriptive, while metalevel prompts (e.g. asking about motivational strategies) seemed harder. When reflective writing is included into a professional development program one should carefully consider the nature and progression of prompts. Concrete situations are the easiest starting point for reflective writing.

4.4 Awareness Activities: Decoding Students' Motivation During Classroom Activities

Teachers' awareness of individual students' motivation is critically important, when they are trying to stimulate motivation within the classroom. Unfortunately, intrinsic motivation is not easily judged from behaviour, as it mingles with other types of motivation. However, Stipek et al. have documented that teachers' capacities to decode students' motivational states can be enhanced if they conduct simple action research in their own classrooms (Stipek et al., 1998). A similar setup may be used to improve teachers' awareness of how specific aspects of the science learning environment shape different students' motivation. We have found two types of awareness activities useful.

4.4.1 Awareness of Individual Students' Motivational State

Inspired by the work of Stipek et al. (Stipek et al., 1998), we asked teachers to make a diverse selection of five students and try to judge their motivation while doing an IBSE-oriented activity. Subsequently, teachers' contrasted their judgments with students' self-reports of motivation. All judgments were registered on the research-based instrument devised and validated by Stipek et al. (23-item Likert scale instrument assessing various motivational dimensions: mastery/performance orientation, perceived ability, persistence, help-seeking strategy, and positive and negative emotions. Four items were omitted, though, since participants found that they would not make sense to their students).

All teachers in our sample severely misjudged at least one of the five students, and one very experienced teacher misread the motivational cues from all target students. The most severe blind spots seemed to be associated with students that differed from teachers in sex, race, attitude, etc. To some extent, all teachers were puzzled by their results, even though some also gained positive insights: "Students report less negative and more positive emotions in relation to chemistry than I assumed… Students' perceived ability and persistence were also higher than I had estimated…". The teachers generally acknowledged the value of both investigation and instrument, and several teachers considered discussing instances of misjudgement with individual students to obtain a better understanding of their motivation and behaviour in the classroom.

4.4.2 Awareness of Motivational Effect of Different Activities

The Intrinsic Motivation Inventory (IMI)[2] is an instrument for investigation of students' motivation in relation to a specific teaching activity. IMI is based on Deci and Ryan's self-determination theory (Deci & Ryan, 2002), and we have provided

---

[2] IMI: http://www.psych.rochester.edu/SDT/measures/IMI_scales.php

teachers with a reduced 28-item version (original 7 scales, only 4 of the original items each). Since validity is assured and since research level reliability is not ultimately important for our purpose, we have not inquired further into the properties of our reduced instrument. Participant teachers collected IMI questionnaire data in relation to a self-selected IBSE activity and subsequently presented their results in a joint workshop session. In general, student scores on interest/enjoyment (measure of intrinsic interest) for the activities were mediocre, though they ascribed value/ usefulness to the tasks. The tasks with the highest intrinsic motivation were physics laboratory tasks that, compared to other tasks, were associated with a pattern of high-perceived autonomy and high-perceived competence. In absolute terms, autonomy was moderate, reflecting that none of the teachers had planned for this aspect. However, the teacher, who consistently (and tacitly) throughout the project gave students most opportunities for independent work, also received the highest autonomy scale scores. The teachers experienced IMI as a useful instrument to map the motivational aspects of different activities, although they questioned the utility of the pressure/tension scale. The indication is that reflection on IMI-like results across different inquiry activities may promote teachers' motivational awareness, e.g. explicating how details in IBSE design influence students' motivation. However, we found that teachers needed software support to make data analysis and reporting more manageable.

4.5   Enacting Motivational Theory and Reflecting in Video Club

Sherin et al. have found that analysis and discussion of video clips from teachers' own classrooms in a video club setting can develop teachers' "professional vision", e.g. their ability to make sense of what is happening in the classroom (Sherin, 2007; Sherin & Van Es, 2009). Video recordings provide a direct enacting-reflecting link, as they permit reviewing of incidents from the classroom supporting teachers' reflection on action. In addition, the classroom videos offer highly contextualized views into other teachers' practices, which might afford vicarious experiences and as such represent an essential source to teacher self-efficacy (Bandura, 1997). Finally, video excerpts provide a common ground for discussion, enabling collaborative scrutiny of motivational dynamics and the building of more complex propositionally interpreted practical knowledge.

Our intervention design relies on video club participation as the most powerful strategy to relate theory and practice. Sherin and others (e.g. Sherin & Han, 2004) have described video clubs as organized contexts for collaborative teacher inquiry into practice through the use of classroom videos. At its core is a group of teachers meeting regularly and over long enough time to develop a sense of community and a responsive culture. They inquire into video clips they in turn provide for discussion, under the guidance of a facilitator that prompts their attention and assists/challenges interpretations of practices. Typically, the facilitator also helps in video club norming and in the practical management of video club discussions. In our context each teacher planned and trialled lessons addressing motivational issues. All trials were videotaped, and teachers selected video excerpts for video club discussion at our joint workshops. Each video should demonstrate how they had worked with certain motivational aspects or illustrating challenging motivational situations. To take into account teachers' autonomy, comfort zones, and concerns about time pressure and local curricula, we did not impose IBSE as a restriction for teachers' trials. Fortuitously, it turned out that all trials on motivational teaching were of an IBSE-oriented nature. This demonstrates that Danish teachers' motivational orientations naturally accommodate IBSE approaches to science teaching, including both practical hands-on activities and more conceptual or discursive activities. In their planned trials, teachers placed emphasis on different motivational constructs, but aspects of competence/self-efficacy, autonomy, relatedness/belonging, and task value were particularly prominent. Below, we present four short cases illustrating how our teachers addressed motivational issues and IBSE approaches.

The four cases illustrate the teachers' different motivational emphases and ways of implementing motivational theories in relation to a range of IBSE-oriented activities. The teachers' presentations in the video club led to essential discussions and reflections on motivational strategies. They also instigated discussions about the challenges and possibilities of "dual purpose teaching" (Palmer, 2005), i.e. teaching simultaneously planned for learning *and* motivation.

Starry Night: students communicate astronomy to a public audience
Motivational focus: autonomy, competence, and relatedness
IBSE aspect: investigating a self-selected topic and authentic communication
The video was recorded on a "Starry Night" where two students from the highly experienced astronomy teachers' class are presenting a topic of their own interest to a public audience of mostly children and their parents. This upper secondary school has such events once every month, and all astronomy students are encouraged to co-plan and present a topic at a Starry Night. Encouragement comes from a teacher-built classroom culture, where communicating to authentic audiences (e.g. primary school students) is part of normal teaching. Further, the teacher consistently raises students' competence beliefs by feedback and signalling: "you can do it!". He also offers scaffolding, whenever students might need it. Preparing for their "Starry Night", students have almost complete autonomy; they themselves decide the astronomic themes to present, seek information and materials for their

(continued)

presentation, and design tasks for participants. At the "Starry Night" event, they direct and teach major parts of the session, with the teacher discreetly backing up; when students signal, they would appreciate it. According to the teacher, "Starry Night" participation stimulates all three innate self-determination needs (competence, autonomy, and relatedness).

The video excerpt presented in video club shows the last part of a student dyad's presentation "Life in the universe", which turned out to be a little shorter than planned. Here the teacher subtly guides into a moderately challenging additional task: why don't you demonstrate inquiry with the IT program Stellarium [astronomy software students have used in ordinary astronomy classes]?

The video club discussions helped identify critical aspects of autonomy and competence and how these constructs are balanced in the "Starry Night" context. Particularly, they added an emphasis on what competence and autonomy supportive teachers actually do. Finally, this video club excerpt initiated considerations of how student experiences of this autonomy supportive kind can be transferred to ordinary science classrooms.

Chemical bonding: constructing a concept map
Motivational focus: relatedness/belonging
IBSE aspect: discursive investigation of concepts and their relationship

The video was recorded in a chemistry class where students were working on a concept map on chemical bonding. The teacher saw most students as interested in chemistry, but acknowledged that some students' motivation might be undermined by their low sense of belonging within the social classroom environment. To address this issue, the teacher had formed groups allowing these marginalized students to interact with core students of the classroom community.

The video excerpt presented in the video club shows a group of three students, including one marginal student. The group members are collaborating on the construction of a concept map of chemical bonding. The video clip reveals how the "marginalized" student is included and makes recognized contributions to the construction of the concept map.

In the video club, it was discussed how teachers can support students' feeling of relatedness and belonging by appropriate design of IBSE tasks and by drawing on knowledge of student group dynamics.

Radioactivity and radon in private homes
Motivational focus: task value, autonomy
IBSE aspect: formulating science questions

The video was recorded in physics class on radioactivity and radioactive half-life. To make this topic relevant for the students, the teacher introduces a

(continued)

newspaper article "Radon in private homes". After in-class reading of the article, students are asked to formulate questions that they would like to know more about. The students are clearly engaged and come up with a lot of questions related to health aspects, which somehow takes the teacher by surprise. He tries to redirect attention to questions that might be answered by physics alone, but students' engagement seems to deteriorate.

The video excerpt presented in the video club shows the students posing questions, the process of sorting questions, and the teachers' handling of a situation with mainly biology-related questions.

In the video club, it was discussed how to handle (proactively, responsively) situations where students' autonomy and choices tend to transcend intended teacher plans and curricula.

Students demonstrating an experiment on chemical reaction rate
Motivational focus: relatedness, autonomy, and competence
IBSE aspect: planning, doing, and presenting experiments
The video was recorded in a chemistry classroom where groups of students should plan and demonstrate an experiment. The teacher gave a brief introduction to the concept of chemical reaction rate, and afterwards students should come up with factors that might influence the rate of reaction and test these experimentally for a specific reaction (metal + acid). The teacher structured an interdependent process, ending with each student group making an upfront demonstration of its findings. The motivational intention was to stimulate students' sense of competence (mastering the experiment), autonomy (design considerations), and relatedness/belonging (working together in groups).

The video clip presented in the video club showed a group of students doing their experiment in front of the class and some students sitting waiting for their turn. The students seemed to be very engaged and eager to demonstrate their experiment.

In the video club, it was discussed as to whether the elements of teacher control of the situation (overall structure, classroom dialogue, emphasis on "scientific methods") could be undermining students' sense of autonomy.

The use of videos and video club participation clearly facilitated development of teachers' motivational thinking and reflection. In the final evaluation, major outcomes were ascribed to watching, analysing, and discussing videos from teachers' own classrooms. All (but one) teachers stated that the video club discussions in general had had a beneficial effect on their teaching. The one teacher with reservations towards discussing videos from other participants' classrooms was an expert chemistry teacher who taught a particular student group. On the other hand, all teachers found individual analysis of videos from their own classroom both interesting and rewarding. As the most prominent outcome, teachers would generally mention that the videos helped them understand what was going on in their classrooms and offered a new perspective to their enactment of teacher roles. Examples of motivational insights would be one teacher realizing how her extensive and deliberate interaction with a couple of low achievers left many other students stuck, waiting for help, and increasingly frustrated. Another teacher suddenly saw how she was undermining students' self-efficacy while trying to convince them into a task they found difficult by assuring them it should be easy. One teacher realized that his understanding of a situation in his classroom differed very much from the understanding held by the other teachers in the video club. Becoming aware of different perspectives and other teachers' interpretations of the whole situation contributed to his understanding of the interaction between him and his students. Another teacher did an excellent job in supporting students' self-efficacy, but he realized that the knowledge he used and his strategies were all tacit. Through video club discussions, he identified motivational elements in his own practice, and his attention was drawn to motivational constructs being activated. Here, the video club discussions contributed to both verbalization and knowledge transformation.

The teacher responses and outcomes we report here are in accordance with targeted studies of professional development through video club participation (Sherin, 2007; Sherin & Van Es, 2009). These studies would recommend that video clubs be run regularly for at least a year, since it takes time to build a supportive video club environment. To enhance sharing, synergy, and propositional interpretation at the video club sessions, it might also be useful to make more focused assignments for teacher trials and video takes than in the above case. Assigning motivational foci like "Supporting or inhibiting students' Self-efficacy" or "Interacting with students' goal-orientations" would still allow teachers considerable autonomy. For analysis and discussion of classroom videos, our CARTAGO framework could serve as an important tool, as it provides structure, guides attention, and enables propositional interpretation of practice.

5   Concluding Remarks

IBSE approaches are at the core of teachers' motivational thinking, and science teachers tend to assume that inquiry approaches automatically will benefit students' motivation. Much of this may be ascribed to a general lack of knowledge of motivational theory among teachers (Fives & Manning, 2005). Furthermore, and against the evidence (Palmer, 2005), many teachers tend to believe that planning for motivation will undermine students' cognitive outcomes. Consequently, much could be achieved if teachers had theoretically informed knowledge about motivation and strategies for incorporating this in their planning and enacting of IBSE. In this chapter, we have argued that this issue may be addressed through in-service teacher training. In support of such efforts, we have tried to address two vital questions: (1) What kind of motivational knowledge and content should we try to engage teachers with? (2) How can we facilitate teachers' transformation and integration of theoretical knowledge about motivation into knowledge for use and practical IBSE teaching?

We have advocated a "pragmatic use" of motivational theory, which implies a focus on a minimum set of educationally relevant, distinct, and functional motivational constructs. We have suggested content for teacher professional development from this pragmatic approach, and we have generated the CARTAGO framework as a heuristic that can be used for planning, analysing, and enacting motivational IBSE teaching. We have briefly indicated how it can be used to pinpoint the motivational weaknesses of traditional laboratory activities (cookbook), but it is just as useful for planning and enacting more motivational IBSE-oriented approaches. Finally, we described three different strategies to interrelate motivational theory and practice that have been applied in our R&D project with upper secondary science teachers. All strategies – reflective writing, awareness exercises (students' motivation), and video clubs – have proven to be fruitful for teachers' professional development. Each strategy explicitly asks teachers to connect theory with practice or vice versa, but they induce different reflecting-enacting aspects, in accordance with the interconnected model of professional development. As such the activities and strategies represent different sources for teacher professional development. A repertoire of strategies is necessary if pervasive and more enduring development of teachers' motivational thinking and practice is the aim. Particularly, video clubs have the potential to support more sustainable enrichment of IBSE practices, as they can proceed locally and beyond the need for intervention. This requires teachers who have appropriated the practices and roles of video clubs and who have had the opportunity to enact these clubs, with an emphasis on motivational and IBSE-oriented teaching.

We do not claim that this is the full story of how to enrich IBSE practices by use of motivational theory, first of all because teachers' orientations towards teaching and traditional teaching practices are deeply anchored and hard to change. In our recent R&D project, we succeeded in enhancing teachers' capacities to analyse and discuss motivational issues, their awareness of students' motivation, and to some extent their planning of motivational IBSE activities, but teachers reported only modest changes towards "more motivational practice". These results indicate that there is a limit to what can realistically be achieved from a relatively short (five workshops, one semester) and nonschool-based in-service training program. This allows us to think that the elements and considerations presented here in essence will survive elaborations to come.

References

Ames, C. (1992). Classrooms – Goals, structures, and student motivation. *Journal of Educational Psychology, 84*(3), 261–271.

Anderman, L. H., & Leake, V. S. (2005). The ABCs of motivation. An alternative framework for teaching preservice teachers about motivation. *Clearing House: A Journal of Education Strategies, Issues and Ideas, 78*, 192–196.

Bandura, A. (1997). *Self-efficacy: The exercise of control*. New York: W.H. Freeman.

Brophy, J. (2004). *Motivating students to learn*. New York: Routledge/Taylor & Francis Group.

Burton, J., Quirke, P., Reichman, C. L., & Peyton, P. K. (2009). *Reflective writing – A way to life-long teacher learning*. TESL-EJ Publications, eBook.

Cherubini, G., Zambelli, F., & Boscolo, P. (2002). Student motivation: An experience of inservice education as a context for professional development of teachers. *Teaching and Teacher Education, 18*, 273–288.

Clarke, D., & Hollingsworth, H. (2002). Elaborating a model of teacher professional growth. *Teaching and Teacher Education, 18*(7), 947–967.

Deci, E. L., & Ryan, R. M. (2002). *Handbook of self-determination research*. Woodbridge, UK: University of Rochester Press.

Fives, H., & Manning, D. K. (2005). *Teachers' strategies for student engagement: Comparing research to demonstrated knowledge*. Paper presented at the annual meeting of the American Psychological Association.

Ford, M. E. (1992). *Motivating humans. Goals, emotions, and personal agency beliefs*. Newburry Park, CA: SAGE Publications.

Martin, A. J. (2008). Enhancing student motivation and engagement: The effects of a multidimensional intervention. *Contemporary Educational Psychology, 33*(2), 239–269.

Nielsen, B. L. (2012). Science teachers' meaning-making when involved in a school-based professional development project. *Journal of Science Teacher Education, 23*, 621–649.

Osborne, J., & Dillon, J. (2008). *Science education in Europe: critical reflections: A report to the Nuffield Foundation*. London: The Nuffield Foundation.

Osborne, J., Simon, S., & Collins, S. (2003). Attitudes towards science: A review of the literature and its implications. *International Journal of Science Education, 25*(9), 1049–1079.

Osborne, J., Simon, S., & Tytler, R. (2009). *Attitudes Towards School Science: An Update*. Paper Presented at the ESERA 2009 conference.

Palmer, D. (2005). A motivational view of constructivist-informed teaching. *International Journal of Science Education, 27*(15), 1853–1881.

Palmer, D. H. (2009). Student interest generated during an inquiry skills lesson. *Journal of Research in Science Teaching, 46*(2), 147–165.

Pintrich, P. R. (2003). A motivational science perspective on the role of student motivation in learning and teaching contexts. *Journal of Educational Psychology, 95*(4), 667–686.

Pintrich, P. R., & Schunk, D. H. (2002). *Motivation in education. Theory, research, and applications* (2nd ed.). Columbus, OH: Merrill Prentice Hall.

Potvin, P., & Hasni, A. (2014). Interest, motivation and attitude towards science and technology at K-12 levels: A systematic review of 12 years of educational research. *Studies in Science Education, 50*(1), 85–129.

Reeve, J. (2002). Self-determination theory applied to educational settings. In E. L. Deci & R. M. Ryan (Eds.), *Handbook of self-determination research* (pp. 183–203). Woodbridge, UK: University of Rochester Press.

Rocard, M., Csermely, D., Jorde, D., Lenzen, D., Walberg-Henriksson, H., & Hemmo, V. (2007). *Science education now – A renewed pedagogy for the future of Europe*. Luxembourg City, Luxembourg: Office for Official Publications of the European Communities.

Sherin, M. G. (2007). The development of teachers' professional vision in video clubs. In R. Goldman, R. Pea, B. Barron, & S. J. Denny (Eds.), *Video research in the learning sciences* (pp. 383–395). Mahwah, NJ: Lawrence Erlbaum Associates, Publishers.

Sherin, M. G., & Han, S. Y. (2004). Teacher learning in the context of a video Club. *Teaching and Teacher Education, 20*(2), 163–183.

Sherin, M. G., & Van Es, E. A. (2009). Effects of video club participation on teachers' professional vision. *Journal of Teacher Education, 60*(1), 20–37.

Shulman, L. S. (1986). Those who understand: Knowledge growth in teaching. *Educational Researcher, 15*(2), 4–14.

Singer, S. R., Hilton, M. L., & Schweingruber, H. A. (2006). *America's Lab report: Investigations in high school science*. Washington, DC: The National Academies Press.

Stipek, D., Givvin, K. B., Salmon, J. M., & Macgyvers, V. L. (1998). Can a teacher intervention improve classroom practices and student motivation in mathematics? *The Journal of Experimental Education, 66*(4), 319–337.

Swarat, S. (2008). What makes a topic interesting? A conceptual and methodological exploration of the underlying dimensions of topic interest. *Electronic Journal of Science Education, 12*(2), 1–25.

Thiessen, A. (2000). A skillful start to a teaching career: A matter of developing impactful behaviors, reflective practices, or professional knowledge? *International Journal of Educational Research, 33*(5), 515–537.

Taking Advantage of the Synergy Between Scientific Literacy Goals, Inquiry-Based Methods and Self-Efficacy to Change Science Teaching

Robert Evans and Jens Dolin

1 Why Scientific Literacy Goals May Not Be Met

The curricular statements of most European countries contain goals for scientific literacies and process competencies. However, we argue that direct targeting by teachers of these goals, or active achievement of them by students, may not actually occur. We report our experience of implementing a teacher development initiative aimed at supporting teachers to target these goals more directly through drawing on the natural relationship between scientific literacy and inquiry-based activities, while also focusing on the support potential of self-efficacy.

Our work is based upon an important review by Roberts (2007) in his contribution to Abell and Lederman (2007) in which he categorizes many international perspectives of scientific literacy (SL) into two visions. Vision 1 includes the products and processes of science from the point of view of the field of science and so can be characterized as the formal knowledge of science. Vision II is contextual and encompasses the intersection of science with the daily lives of learners and so represents the societal relevance of science (Roberts, 2007). These complementary views are combined in various ways in most statements of SL goals as can be seen in the following examples derived from national policy statements for Denmark and Scotland (S-Team, 2012):

Examples of Vision I statements

A. Students can carry out observations in laboratory to establish simple hypotheses. (Denmark)
B. Students use Information Communication Technology as a tool in processing data. (Denmark)

R. Evans (✉) · J. Dolin
Department of Science Education, University of Copenhagen, Copenhagen, Denmark
e-mail: evans@ind.ku.dk; dolin@ind.ku.dk

© Springer International Publishing AG, part of Springer Nature 2018　　　　105
O. E. Tsivitanidou et al. (eds.), *Professional Development for Inquiry-Based Science Teaching and Learning*, Contributions from Science Education Research 5, https://doi.org/10.1007/978-3-319-91406-0_6

Examples of Vision II statements

C. The scientifically literate person reflects critically about information in the media and reports. (Scotland)
D. The scientifically literate person develops informed social views about the environment and other issues. (Scotland)

When preparing to teach for these goals, experience shows that not only are they often difficult to achieve, but traditional teaching methods may be inadequate for reaching them. For example, goal 'B' for Vision I might typically be approached by using teacher demonstrations of the use of Information Communication Technology followed by student practice with given problem sets. However, without deeper interaction through personal experience with the decisions associated with selecting data processing strategies relevant to unique and authentic situations, this Vision I SL goal will not be fully met. Similarly, teaching aimed at a societally relevant SL goal from Vision II may not be successful when it is decontextualized. Students may learn only the '...correct professional concepts' instead of actual immersion in a local issue in meeting the Vision II example in 'D' above. Students may even be tested on definitions or even simulated cases with these concepts but not be able to achieve the actual Vision II SL objective of contextual use of science for approaching authentic societal problems.

Consequently, our research questions were:

– Can concept networks of national scientific literacy goals that reveal the products and processes of Robert's (2007) Vision I and II be matched with inquiry teaching and learning methods to help teachers address them in their teaching?
– Can teacher self-efficacies for using inquiry methods connected to scientific literacy purposely be raised to increase the likelihood of further inquiry use?

2   Features of Inquiry-Based Teaching and Learning and Self-Efficacy

Inquiry-based science teaching and learning (IBST/L) is an umbrella term, covering a wide spectrum of teaching approaches, all of which place the students' own questions as the key drivers for constructing new knowledge (Spronken-Smith, Walker, Batchelor, O'Steen, & Angelo, 2012). In many ways it can be seen as a realization of the principles of constructivism (Llewellyn, 2007). It has many different concrete designs and formulations, all of which focus on specific features characterizing science, such as addressing problems, performing investigations, searching for information, constructing models, debating with peers, etc.

The perspective of inquiry teaching and learning which we chose to guide our work in facilitating inquiry-based teaching methods has been summarized by NRC

(2000) (refer to Fig. 2 in Chapter "Introduction: What Is Inquiry-Based Science Teaching and Learning?") as well as the EU project *Mind the Gap* (MTG, 2007) in the following four essential parts:

(i) Authentic and problem-based learning activities where there may not be a correct answer
(ii) A certain quantity of experimental procedures, experiments and 'hands-on' activities, including searching for information
(iii) Self-regulated learning sequences where student autonomy is emphasized
(iv) Discursive argumentation and communication with peers ('talking science')

These four dimensions link inquiry-based science teaching and learning to a specific educational set-up:

(i) Problem-based learning – to emulate the scientific processes and to promote the development of generic competences
(ii) Experiments and 'hands-on' activities – one of the key features of science and science education
(iii) Student autonomy and active involvement – to motivate and engage students in learning processes and to develop their personal and generic learning competences
(iv) Learning through dialogue and seeing argumentation as the link between the real world and scientific understanding of the world

Our contention in designing teacher professional development materials was that inquiry-based teaching, which is especially useful in preparing students for problem-solving and process skill acquisition, is an ideal methodological match for the demands of teaching for scientific literacy. This is because it immerses students in the precise processes of scientific work and thinking that are often called for in both Vision I and II goals.

For our work with SL and IBST/L, we also employed the natural relationship between self-efficacy and teaching studies which show that teaching behaviours such as persistence at tasks, risk-taking and the use of innovations like IBST/L are all related to levels of self-efficacy (see Ashton & Webb, 1986; Czerniak, 1990; Enochs & Riggs, 1990; Guskey, 1988; Gibson & Dembo, 1984; Woolfolk & Hoy, 1990). For example, in science teaching, teachers with high self-efficacies were found to be more likely to use inquiry and student centred teaching methods while those with low efficacies were more likely to be teacher directed (Czerniak, 1990). Research has also shown that teacher self-efficacy beliefs strongly influence the nature of a teacher's role, planning and consequently curriculum and student learning (Tobin, Tippins, & Gallard, 1994).

In 1997, Albert Bandura suggested detailed mechanisms by which the self-efficacy of teachers can be maintained, raised or diminished. Significantly, he strengthened the link between self-efficacy and the extent to which it is influenced by the context of the situations in which teaching is performed. He further differentiated self-efficacy from other less malleable constructs such as self-confidence and self-concept which are both more general and less situation specific (Bandura, 1997). Bandura's proposed mechanisms by which teacher self-efficacy may be influenced are:

1. Mastery of teaching experiences
2. Modelling of other teachers
3. Authentic and valid performance feedback
4. Environments where stress is not inhibiting

Consequently, we developed a teacher professional development workshop which consciously included all of these strategies. In our workshop, mastery experiences were achieved through collaborative lesson transformations from traditional to inquiry-oriented. Videos of microteaching experiences that included supportive peer feedback and inquiry lessons addressing SL goals were shared with other participants. In these same experiences, teacher educators and teachers were able to observe colleagues teaching and so model their own teaching on the success of others. They also got and gave supportive and valid performance feedback, which helped increase individual capacity beliefs. Since the environments of the workshops for teacher professional development were consciously non-threatening and positive in supporting risk-taking and growth, overall stress at using somewhat difficult inquiry methods with challenging SL goals was reduced.

To summarize, our perspective in designing teacher professional development materials was that inquiry-based teaching is especially useful in preparing students for problem-solving and process skill acquisition and therefore is an ideal methodological match for the demands of teaching for scientific literacy. This is because it immerses students in the precise processes of scientific work and thinking that are often called for in both Vision I and II goals. For our work with SL and IBST/L, we also employed the natural relationship between self-efficacy and teaching studies that show that teaching behaviours such as persistence at tasks, risk-taking and the use of innovations like IBST/L are all related to levels of self-efficacy.

3   Teachers Need Conceptual Tools to Navigate This Synergy

To actively address the common problem of reaching scientific literacy goals, we developed a teacher professional development workshop that advocates IBST/L methods as a means to achieve such goals and actively enhanced teacher self-efficacies, so that participants would be more likely to subsequently use the inquiry methods. The workshop was grounded in two conceptual tools. The first used concept networks to reveal the contents of national science literacy statements, and the second transformed standard lessons to IBST/L formats. Since targeted and specific self-efficacy beliefs increase the success and persistence of unfamiliar teaching methods (Mulholland & Wallace, 2001), we also used Bandura's (1997) concept of higher self-efficacies targeted at specific future behaviours as helpful beliefs when working with challenging teaching methods such as IBST/L.

Teachers first need to be aware of and fully understand their national goals for scientific literacy. We found across the five cultures where we conducted workshops that teachers were at the most vaguely aware of their SL goals. Even when they were presented with them, the abstractions of the textual sentences were difficult to grasp. Therefore, we used the conceptual tool of graphical representation via networks of the goals to reduce these textual abstractions. With this heightened understanding of SL goals, we then helped teachers connect SL process of science objectives to inquiry teaching where science process methods have a significant role.

The workshops were themselves based on the potential of IBST/L (see Chapter "Introduction: What Is Inquiry-Based Science Teaching and Learning?") and therefore used IBST/L methods to model the methods being advocated. They consisted of:

- Group experience with an inquiry activity followed by deconstruction of the activity to determine the component parts (NRC, 1996)
- Analyses of short science teaching videos for both positive examples of inquiry instruction as well as missed opportunities for inquiry teaching
- Examination and comparison of national statements of scientific literacy in order to identify learning objectives for workshop teaching
- Transformation of traditional science lessons into IBST/L lessons (Schroeder, Scott, Tolson, Huang, & Lee, 2007)
- Microteaching invitations to inquiry for the workshop group
- Subsequent supportive feedback of these short inquiry lessons
- Full class IBST/L aimed at scientific literacy goals in each participant teacher's own classes
- Sharing video segments from participants' own classroom IBST/L trials for group feedback at a follow-up meeting
- Promotion of participants' positive self-efficacies during each of part of the workshop (Bandura, 1995)

We iteratively tested trial versions of this workshop in Scotland, Denmark, Norway, Hungary and Egypt. Each time we collected open-ended answers to standard questions concerning the strengths and weaknesses of the workshop, along with quantitative assessments of participant self-efficacy taken two or three times during each trial. Although the groups for these five trials were small and we did not collect demographics about them, the diversity of cultures and consistent feedback from the five groups allow us to claim some generalizable outcomes. Analysis of the open-ended feedback allowed us to identify elements of the workshop that were useful and to address any aspects that were not. Given the perspectives of five national teaching cultures, we were able to identify and keep activities found to be relevant and helpful to a cross-national population. For example, in our first trials of the workshop, we asked participants to create inquiry lessons in small groups, video record them and present them in the workshop. We received a number of suggestions about this activity, similar to this one: *'Let teachers go home to their own classrooms and let them video one another. Then come together again to discuss the videos. Much less 'artificial' videos this way. Interesting to see how teachers who agree about methods in a workshop actually put them into practice.* In response to these comments, we now ask the teachers to prepare and teach short IBST/L micro-lessons during the workshop and then to return to their classrooms to teach a full IBST/L lesson which they record and share with participants at a later workshop meeting. Other suggestions from the open-ended feedback were also iteratively incorporated into our workshop. Consequently, the workshop details explained here are summative, based on each successive group's experiences, and therefore were not tested on all five groups.

In his original work, Bandura (1977) showed that based on life experiences, people have specific expectancies about their action-outcome contingencies. They act not only because they believe their actions will result in specific outcomes but also because they believe in their own ability to perform those actions. We therefore hypothesized a mechanism with which teaching environments, such as the use of IBST/L methods to reach goals of scientific literacy, can interact with and modify teachers' beliefs about their science teaching self-efficacy and consequently the quality of science teaching and learning (Andersen, Dragsted, Evans, & Sørensen, 2005). Since self-efficacy may be related to successful science teaching (e.g. Bandura, 1997; Dembo & Gibson, 1985; Tschannen-Moran, Hoy, & Hoy, 1998), changes in self-efficacy may be useful in helping teachers of science become more successful.

4   Using Conceptual Tool 1: Concept Networks of National Standards

Our work with science teacher educators and science teachers in five countries showed us that many were either unaware of their national scientific literacy statements or did not consider them when planning for teaching. We hypothesized that one obstacle to using these curricular objectives might be difficulties with identifying and understanding the goals in relation to more traditional goals and then recognizing their potential usefulness. Thus, we transformed national statements into

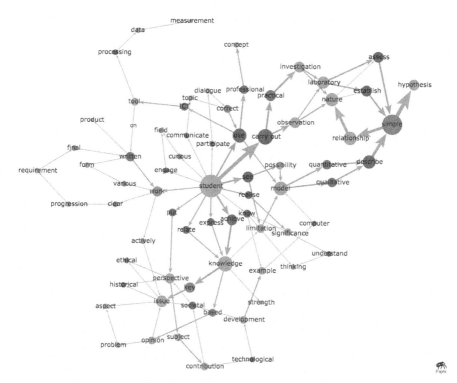

Fig. 1 A Danish 'weighted' network of scientific literacy concepts derived from Danish national statements. The size of both circles and connecting lines is relevant to the frequency of their use in the national statements, as are their positions from the centre towards the edges of the map. As an illustration, one path example can be seen by reading from the starting point at 'student' to 'carry out' and on to 'observations' in 'nature' (see Bruun, Dolin and Evans, 2009)

concept networks that allowed SL to be more appropriable. Using network analysis algorithms originally developed for physics by de Nooy, Mrvar, and Batagelj (2011), we constructed networks that visually link, by both significance and relevance, the various statements in national SL documents.

The networks have been produced with PAJEK (Batagelj & Mrvar, 1996, 2007) software based on complex network theory. As a special feature, the networks reveal strings of defining elements (concepts, actions, contexts, levels, etc.) showing the relative importance of the connections between the elements. The point is that such networks make up a visual representation of often complex texts with an integrated quantitative approach. A detailed presentation of the process can be found in Bruun, Dolin and Evans (2009).

As examples, Figs. 1 and 2 show concept networks of Danish and Scottish SL curricular statements. They can be read by starting with the 'actor', which is a 'student' in the Danish map and a 'scientifically literate person' in the Scottish sche-

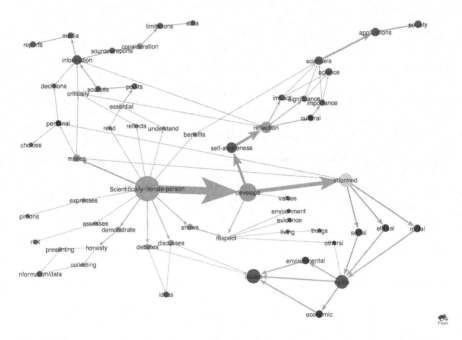

Fig. 2 A Scottish 'weighted' network of scientific literacy concepts derived from Scottish national statements. The size of both circles and connecting lines is relevant to the frequency of their use in the national statements, as are their positions from the centre towards the edges of the map. In this example, the starting point is 'scientifically literate person' and statements then follow from 'develops' to 'informed' 'social' 'views'

matic. These actors can then perform a number of 'actions' such as 'carry out', 'use', 'reflect' and 'develop', as seen in the Vision I and Vision II examples given earlier.

When SL networks are used with science teacher educators and science teachers, they first need to understand how the networks are constructed from the original written SL statements and then have some time to explore and discuss the maps in order to see how they represent not only the content of the text but the less apparent interrelationships and relative emphases of various aspects of Vision I and Vision II perspectives of scientific literacy. Throughout the project, the scientific literacy networks appeared to improve the accessibility and comparability of national statements of scientific literacy based on the ability of workshop participants to analyse and discuss scientific literacy when using them rather than when using textual statements. The ability to actually see where the emphases are placed proved to be a stimulating and useful perspective in the use of these materials. We validated this use of networks for teachers through multiple national trials and in each case, the networks proved useful for improving the accessibility of SL for inquiry-oriented workshops where participants were actively engaged in constructing understanding rather than consuming textual statements (S-Team, 2012).

5   Using Conceptual Tool 2: Transforming Traditional Lessons to Inquiry

We also found it useful for participants in our teacher professional development workshops to transform good traditional science 'labs', which merely relate to SL goals, to increasingly inquiry-oriented versions which can more nearly achieve a given SL objective. These transformations made the changes concrete and visual, particularly in the table format that allowed for easy comparisons. For example, methods for teaching the science lesson that follows can be conceptualized as either relatively traditional or based on the *Mind the Gap* caveats for inquiry (Dolin, Evans, & Bruun, 2010; MTG 2007) and the guidelines from Loucks-Horsley and Olson (2000) for inquiry teaching and learning, more inquiry-oriented.

The following example of transitioning a lesson (see Table 1) was used to show how to help pupils achieve the Danish Vision I objective A: 'Students can carry out observations in laboratory to establish simple hypotheses'. It is based on the following science content.

Using Chromatography to Examine Plant Pigments
When plant leaf cells are crushed and mixed with a solvent, their component pigments (primarily their chlorophylls, beta carotene and xanthophyll) can be separated from one another using paper chromatography. This suspension is placed near one end of a strip of chromatography paper and the other end is placed in the solvent. By capillary action, the solvent moves up the paper taking the pigments along with it. However, since each of the pigments has a different molecular structure, they move different distances. Chlorophylls move the shortest distance because their chemical bonds to the paper are stronger than the other pigments. Xanthophyll forms hydrogen bonds with the paper and is not very soluble, so it does not travel too far, but farther than the chlorophylls. Beta-carotene is carried the furthest due to its high solubility and lack of hydrogen bonding with the paper.

Table 1 Progressive variations of a traditional science activity towards inquiry (left to right) and to a closer alignment (also left to right) with scientific literacy objective Vision I A from Denmark: 'Students can carry out observations in laboratory to establish simple hypotheses' (Denmark)

Essential features of a Vision I scientific literacy goal A for Denmark (below)	Variations			
	Progressive transitions from more traditionally oriented lessons from just below in the left column to more inquiry-oriented lessons in the right-hand column in order to meet the scientific literacy goal A for Denmark (the far left column)			
Student engages in scientifically oriented questions	Can standard paper chromatography be *used to* validate teacher suggestion that it can be used to separate the four pigments of plant cells?	Can standard paper chromatography be used *to find out* how many plant pigments are in a cell?	Can plant cell pigments and their characteristics be *explored* and *described* in unique ways using paper chromatography?	Can student *devised* paper chromatography techniques and questions be used to *provide clues* to the differences in plant pigments including something about their molecular structure?
Student gives priority to evidence in responding to questions	Student *uses* given lab instructions to separate out the four pigments	Student *uses* lab instructions to *discover* the number of pigments	Student *devises* several ways to use chromatography to reveal pigment characteristics	Student *perfects* own methods for using chromatography to *discover* pigments characteristics and *analyses* the results for clues to their molecular structures
Student formulates explanations from evidence	Students *report* finding all four pigments	Students *explain* and *describe* different pigments on the basis of the distance they travelled	Students *hypothesize* about the causes of pigment differences based on their distance travelled	Students use different travel distances and paper molecular structure to *hypothesize* about pigment structures
Student connects explanations to scientific knowledge	Student *checks* results against those which were given in the lab	Students *confirm* their descriptions with 'expert' knowledge of pigment chromatography	Students *rationalize* their hypotheses on the basis of different sizes of pigments	Student *applies* knowledge of hydrogen-oxygen bonding to explain different pigment distances moved
Student communicates and justifies explanations	Student *completes* given tables of results and answers given questions	Student *writes* a 'lab report' based on collected data	Students *explain* their differing hypotheses and *defend* them orally	Students use molecular models to *explain* their hypotheses to others and *defend* them with the data collected

The lesson encourages joint productive activity	Students can work *singly* or in *groups* to follow the given directions and complete the given tables	Students can work *singly* or in *groups* to find out about pigments and calculate their movement	Students work in *cooperative groups* to explore, describe and share observations about pigments	Students work in *cooperative groups to devise* and share methods and hypotheses
The lesson promotes curiosity and interest	Some interest in chromatography procedures and results	Interest in both chromatography and in *finding* out the correct number of pigments in plant cells	Interest in chromatography and in the challenge of *hypothesizing* about reasons for different distances	Interest in materials, challenge of *devising methods* and curiosity in how such a macroscopic procedure could give insights into molecular characteristics
The lesson provides choice and control	Students *follow closed lab* directions for set-up and fill in given tables with results	Students *follow more open lab* directions for set-up and fill in given tables with results	Students devise methods and hypotheses	Students devise own methods for testing and analysis
The lesson stimulates cognitive engagement and challenge	Students may be challenged to *follow* lab procedures with precision since results will vary with various techniques	Students may be challenged to follow lab procedures with precision and will be stimulated to correctly *find* the number of pigments	Students will be engaged to *create* and try various methods of running the chromatography and to create hypotheses to explain the results	Students will be stimulated *to figure out* the best way to run the lab and challenged to make sense of it at a molecular level based on the results
Creates personal relevance	*Understanding* more about chromatography they may have experienced with ink stains	Understanding chromatography from own life and *solving* a problem using a new tool	Getting to try on the role of a scientist by *figuring out* the best procedure and 'experimenting'	Understanding chromatography, getting to feel like a scientific team with a *challenging problem*
The lesson supports dialogue, literacy and/or research skills	Work and/or report in groups based on a careful reading of the lab directions	Work *and/or report in groups* which validate their results from sources	Work in *cooperative groups* and depend on established research to validate their results	Work in cooperative groups and use high-level sources to help make sense of results

Adapted from Loucks-Horsley and Olson (2000, p. 29) and applied to SL

6 Results of Practical Application of the Tools

Both concept networks of national statements (Conceptual Tool 1) and transforming traditional lessons towards inquiry lessons (Conceptual Tool 2) helped teachers appreciate the synergy between IBST/L methods and scientific literacy goals.

In analysing the concept networks of scientific literacy (SL) from five European countries, we found that many statements of scientific literacy not only allow for but also may best be met with inquiry-based science teaching to fully achieve their goals. The realization that IBST/L teaching and SL complement each other turned out to be both motivating and useful in our teacher professional development workshops.

For example, in the Danish literacy map (see Fig. 1), a Vision I goal states that students should '… carry out practical investigations in the laboratory to establish simple hypotheses and to assess simple hypotheses'. Inquiry-based science teaching may be the best pedagogical method to achieve this goal since through the constructive engagement of the student in an inquiry laboratory, these higher-level goals can be achieved. A traditional transmissive mode of teaching, where students follow standard 'labs' where they are told what to do to in order to 'prove' a given hypothesis, makes it difficult to achieve the goal of allowing students to 'establish' and 'assess' simple hypotheses themselves (Schroeder et al. 2007). They could only confirm those given to them by their teachers and textbooks. In our workshops, we all realized that some of the frustrations of meeting constructivist SL goal statements could be reduced through IBST/L.

Similarly, in the Scottish statements, Vision II goals such as 'The scientifically literate person reflects critically about information in the media and reports' may be difficult to achieve with transmissive teaching while it is easier in inquiry environments where students are engaged to think creatively and deeply. In order for pupils to genuinely achieve such goals, they must do more than follow teacher or textual suggestions for reflection; they need to use scientific concepts and understanding in active forms of personal or public discourse. Well-led teacher lessons can provide stimuli for such reflections. Moreover, teacher facilitation, a useful method in IBST/L teaching, is more supportive of student constructivism than more teacher-dominated methods. Teacher participants in our workshops concurred with the appropriate fit between inquiry methods and such SL statements.

Creating gradations of inquiry such as in Table 1 helped the teachers in our workshops transform lessons towards more use of inquiry methods and to actively construct an understanding of how inquiry methods can align with Vision I and Vision II SL objectives. They concluded that not all SL goals should be taught using the full inquiry methods of the right-hand column. Rather, they decided that each SL goal should be evaluated to determine at which point on the continuum in Table 1 it could be positioned to better meet each goal.

We can trace one example (due to space) of changes in teacher thinking from their understanding of a pathway in a concept network through a transition to a more inquiry-based lesson. A group of Danish teachers examined the national concept network and noted that a goal of Danish SL, 'The student can carry out practical investigations in the laboratory to establish simple hypotheses', is essential but not quite in alignment with their current practices. Comparing the actions of the goal ('carry out' and 'establish') to how they were currently using lab exercises, they saw that while they help students 'carry out' a laboratory exercise by following teacher directions and 'establish' hypotheses through prompting worksheet questions, such lessons did not align with the active student nature of the goal. Consequently, the teachers transformed their traditional lessons along the continuum seen in Table 1 towards IBST/L. Instead of telling their students how to use paper chromatography and then giving them templates for recording their data and finally leading them with worksheet questions to arrive at hypotheses to explain the data, the students devised their own methods for making measurements, ways of representing their data and creating their own hypotheses. After overcoming difficulties in motivating students accustomed to being given directions for labs, the teachers were themselves motivated by early signs of authentic student engagement in scientific practice and hence scientific literacy. These early efforts and the group support they received from fellow teachers began a promising increase in self-efficacy for future transformations.

The self-efficacy measure (modified from Bleicher, 2004; Enochs & Riggs, 1990) given at the start and conclusion of the workshops provided evidence for both the direction and extent of changes in teacher capacity beliefs. For example, post-workshop self-efficacy beliefs with 11 teachers increased over pre-workshop levels. While not a significant difference, these positive trends in self-efficacy indicated that the participants may have felt better able to teach goals for scientific literacy using IBST/L methods than they did before the workshop. Our observationally based beliefs are that the repeated teaching trials beginning in the 'safety' of the workshop and continuing in participant's classrooms provided self-efficacy increasing 'mastery' experiences. In addition, the workshop teaching experiences included authentic peer and course leader feedback, all of which enhance self-efficacies Bandura (1997). Consequently, we have increased our active efforts to influence self-efficacy and are now working to discover whether we can connect specific actions on our part with changes in self-efficacy.

Through the iterative testing of the workshop across five countries, we adjusted teachers' expectations to the difficulties in challenging students to harder 'work' with IBST/L and the high value of peer feedback from colleagues also engaged in transformations.

7   Conclusion

The concept networks and the transformation tools at first sight look complex. However, with the examples of transformation in Table 1 from the Danish group of teachers, we have illustrated how teachers can use them to navigate the synergy between SL goals and inquiry methods while developing self-efficacies likely to sustain their future efforts. The methodology in our workshops supports teachers in being motivated and feeling they have self-efficacy in moving towards inquiry methods and in knowing how far to move in this direction for particular SL goals. These teaching moves towards inquiry and increased self-efficacy for teachers are also likely to enhance student confidence and motivation to handle the extra work required of them in IBST/L. We have developed a framework for supporting SL through inquiry that is both practical (and tested as such across five countries) and theoretically coherent and so should interest teachers, teacher educators as well as supporting educational research.

Our teacher professional development workshop facilitated the use of IBST/L methods to achieve scientific literacy goals and was grounded in the active enhancement of the personal agency belief of self-efficacy. The basic strategy was that by empowering teacher educators and teachers to work towards more motivating SL goals using the kind of inquiry described by MTG (2007), we could actively enhance their self-efficacy for new science teaching more readily than by using traditional teaching methods with less motivating content (see Andersen, Dragsted, Evans, & Sørensen, 2004) for a more complete expression of the theoretical bases.

Teachers in our workshops in Denmark, Scotland, Norway, Hungary and Egypt indicated that achieving SL goals can be challenging, particularly those of Vision II which may be interdisciplinary and include societal connections with science which many teachers feel unprepared to address. Using inquiry methods is also difficult for many teachers due both to their lack of training and the inherent difficulties in using these methods. However, since teaching self-efficacy is associated with the likelihood of using inquiry methods (Czerniak, 1990) and since SL goals lend themselves to inquiry methods where teachers can facilitate student interaction with Vision I goals of science as well as with the Vision II societal implications, increasing self-efficacies may be a useful strategy for achieving SL objectives through inquiry methods.

In summary, this chapter reports our experience using the natural relationship between scientific literacy and IBST/L, as well as self-efficacy, in teacher professional development. It suggests that teacher educators use concept networks of national scientific literacy statements to help science teachers readily identify authentic objectives and then design inquiry-based lessons to achieve them. To facilitate this change of traditional science lessons, we provide an example of such a stepwise transition towards a more inquiry-oriented teaching. Our contention in designing teacher professional development materials is that IBST/L, which is especially useful in preparing students for problem-solving and process skill acquisition, is an ideal methodological match for the demands of teaching for scientific literacy. This is because it immerses students in the same processes of scientific work and thinking that are called for in scientific literacy.

We also give indications that using IBST/L methods to achieve scientific literacy goals can provide important motivation for students and teachers. Students can be motivated to engage in scientific processes and teachers in changing their teaching towards more inquiry-based methods.

References

Abell, S., & Lederman, N. (2007). *Handbook of research on science education.* London: Erlbaum.

Andersen, A., Dragsted, S., Evans, R., & Sørensen, H. (2005). *The Relationship of Capability Beliefs and Teaching Environments of New Danish Elementary Teachers of Science to Teaching Success.* Paper presented at the European Science Education Research Association Conference, Barcelona, Spain, August 2005.

Andersen, A. M., Dragsted, S., Evans, R. H., & Sørensen, H. (2004). The relationship between changes in teachers' self-efficacy beliefs and the science teaching environment of Danish first-year elementary teachers. *Journal of Science Teacher Education, 15*(1), 25–38.

Ashton, P. T., & Webb, R. B. (1986). *Making a difference: Teachers' sense of efficacy and student achievement.* New York: Longman Publishing Group.

Bandura, A. (1977). Self-efficacy: Toward a unifying theory of behavioral change. *Psychological Review, 84*(2), 191–215.

Bandura, A. (1995). *Self-efficacy in changing societies.* New York: Cambridge university press.

Bandura, A. (1997). *Self-efficacy: The exercise of control.* New York: W.H. Freeman and Company.

Batagelj, V., & Mrvar, A. (1996, 2007). *Pajek: Program for analysis and visualization of large networks.* http://pajek.imfm.si/doku.php?id=download

Bleicher, R. E. (2004). Revisiting the STEBI-B: Measuring self-efficacy in preservice elementary teachers. *School Science and Mathematics, 104*(8), 383–391.

Bruun, J., Dolin, J., & Evans, R. (2009). *Diversity of scientific literacy in Europe.* Paper presented at ESERA2009, Istanbul.

Czerniak, C. M. (1990). *A study of self-efficacy, anxiety, and science knowledge in preservice elementary teachers.* Atlanta, GA: National Association for Research in Science Teaching.

De Nooy, W., Mrvar, A., & Batagelj, V. (2011). *Exploratory social network analysis with Pajek* (Vol. 27). New York: Cambridge University Press.

Dembo, M. H., & Gibson, S. (1985). Teachers' sense of efficacy: An important factor in school improvement. *The Elementary School Journal, 86*(2), 173–184.

Dolin, J., Evans, R., & Bruun, J. (2010). *Report on understanding of IBST and Scientific Literacy,* In University of Oslo. Mind The Gap: Seventh Framework Programme, Commission of the European Communities (Grant Agreement # 217725).

Enochs, L. G., & Riggs, I. M. (1990). Further development of an elementary science teaching efficacy belief instrument: A preservice elementary scale. *School Science and Mathematics, 90*(8), 694–706.

Gibson, S., & Dembo, M. H. (1984). Teacher efficacy: A construct validation. *Journal of Educational Psychology, 76*(4), 569.

Guskey, T. R. (1988). Teacher efficacy, self-concept, and attitudes toward the implementation of instructional innovation. *Teaching and Teacher Education, 4*(1), 63–69.

Llewellyn, D. (2007). *Inquiry within: Implementing inquiry-based science standards in Grades 3–8.* Thousand Oaks, CA: Corwin Press.

Loucks-Horsley, S., & Olson, S. (2000). *Inquiry and the national science education standards: A guide for teaching and learning.* Washington, DC: National Academy Press.

MTG. (2007). *Mind the Gap: Learning, Teaching, Research and Policy in Inquiry-Based Science Education.* Grant Agreement Number 217725, European Commission.

Mulholland, J., & Wallace, J. (2001). Teacher induction and elementary science teaching: Enhancing self-efficacy. *Teaching and Teacher Education, 17*(2), 243–261.

National Research Council. (1996). *National science education standards.* Washington, DC: National Academy Press.

National Research Council. (2000). *Inquiry and the national science education standards.* Washington, DC: National Academy Press.

Roberts, D. (2007). Scientific literacy/science literacy. In A. K. Abell & N. G. Lederman (Eds.), *Handbook of research on science education* (pp. 729–780). Mahwah, NJ: Lawrence Erlbaum Associates, Publishers.

Schroeder, C. M., Scott, T. P., Tolson, H., Huang, T. Y., & Lee, Y. H. (2007). A meta-analysis of national research: Effects of teaching strategies on student achievement in science in the United States. *Journal of Research in Science Teaching, 44*(10), 1436–1460.

Spronken-Smith, R., Walker, R., Batchelor, J., O'Steen, B., & Angelo, T. (2012). Evaluating student perceptions of learning processes and intended learning outcomes under inquiry approaches. *Assessment & Evaluation in Higher Education, 37*(1), 57–72.

S-Team. (2012). Final Report for FP7, SiS 2008, action 2.2.1.1.

Tobin, K., Tippins, D. J., & Gallard, A. J. (1994). Research on instructional strategies for teaching science. In D. L. Gabel (Ed.), *Handbook of research on science teaching and learning* (pp. 45–93). New York: Macmillan Publishing Company.

Tschannen-Moran, M., Hoy, A. W., & Hoy, W. K. (1998). Teacher efficacy: Its meaning and measure. *Review of Educational Research, 68*(2), 202–248.

Woolfolk, A. E., & Hoy, W. K. (1990). Prospective teachers' sense of efficacy and beliefs about control. *Journal of Educational Psychology, 82*(1), 81.

Inquiry-Based Approaches in Primary Science Teacher Education

Sami Lehesvuori, Ilkka Ratinen, Josephine Moate, and Jouni Viiri

1 Introduction

The essential features of classroom inquiry have been described as follows:

- Learners are engaged by scientifically oriented questions.
- Learners give priority to evidence, which allows them to develop and evaluate explanations that address scientifically oriented questions.
- Learners formulate explanations from evidence to address scientifically oriented questions.
- Learners evaluate their explanations in light of alternative explanations, particularly those reflecting scientific understanding.
- Learners communicate and justify their proposed explanations (National Research Council, 2000. p. 25).

It can be summarized that it is important for students to consider their own ideas and arguments alongside experimental exercises. According to this characterization, communication and students' preconceptions are both important features in inquiry-based teaching. As social aspect has been neglected in IBST (Oliveira, 2009), in this chapter, this is addressed through emphasis on classroom interaction and its visualizations.

This chapter introduces an interactional graphic tool (see Figs. 1 and 2) for science-based classroom interaction used to map, reflect and interpret inquiry-based science classroom interaction. The interactional graphic offered a means to map the communicative approaches used by the student teachers when teaching. Mapping the communicative approaches of the student teachers provided a visual record of

S. Lehesvuori (✉) · I. Ratinen · J. Moate · J. Viiri
University of Jyväskylä, Jyväskylä, Finland
e-mail: sami.lehesvuori@jyu.fi; ilkka.ratinen@jyu.fi; josephine.moate@jyu.fi; jouni.viiri@jyu.fi

© Springer International Publishing AG, part of Springer Nature 2018
O. E. Tsivitanidou et al. (eds.), *Professional Development for Inquiry-Based Science Teaching and Learning*, Contributions from Science Education Research 5, https://doi.org/10.1007/978-3-319-91406-0_7

121

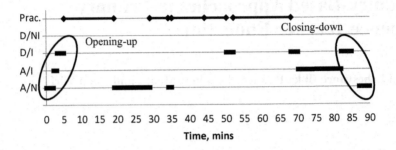

Fig. 1 Lesson diagram of Class A (*A/NI* authoritative and noninteractive, *A/I* authoritative and interactive, *D/I* dialogic and interactive, *D/NI* dialogic and noninteractive, *Prac.* practicing phase)

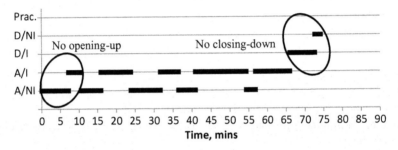

Fig. 2 Lesson diagram of Class B (*A/NI* authoritative and noninteractive, *A/I* authoritative and interactive, *D/I* dialogic and interactive, *D/NI* dialogic and noninteractive, *Prac.* practicing phase)

the interactional patterns employed in the science lessons. As classroom interaction can be analysed and interpreted in many ways, visualization addresses the temporality of interaction (Lehesvuori, Viiri, Rasku-Puttonen, Moate, & Helaakoski, 2013). Visual illustration provides a useful tool for both reflecting on classroom interaction and identifying areas for further development, applicable to educational research as well as initial and in-service teacher development. In this study, we demonstrate how visualizations can be used to evaluate IBST.

The main concepts drawn upon for our theoretical framework are inquiry-based learning, dialogic teaching and the communicative approach (Mortimer & Scott, 2003). It is interesting to note that while inquiry-based approaches tend to overlook the dialogic aspect of science teaching and learning, the main principles of inquiry-based approaches are, in many ways, related to the fundamental ideas of dialogic teaching. The depth of understanding offered by dialogic and communicative approaches concerning the complex interactions of science teaching and learning complements an inquiry-based approach. Furthermore, including dialogic and communicative approaches in science education addresses concerns about the lack of openness in inquiry-based approaches. This means that learners are working towards predetermined outcomes, as if they were following a prescribed recipe (Sadeh & Zion, 2009). Authenticity and openness cannot be conveyed only through applying transmissive and authoritative forms of interaction.

A fundamental aim of dialogic teaching is to explicitly extend pupil reasoning and understanding. Pupil activity is essential in dialogic teaching. The key characteristics of a dialogic approach (Alexander, 2006) can be briefly described as being:

- Collective: Teacher and pupils jointly participate in the learning as a group or as a class.
- Reciprocal: Teacher and pupils listen to each other, share ideas and consider alternative views.
- Supportive: Pupils can present their ideas freely without fear of being incorrect.
- Cumulative: Teacher and pupils develop their ideas together, jointly constructing knowledge.
- Purposeful: The teacher plans and guides the discourse, paying attention to educational goals in addition to the above points.

Of these characteristics, the aim of developing ideas cumulatively is particularly difficult to achieve (Alexander, 2006) and requires that the teacher has high-quality professional skills including genuine subject knowledge, appropriate pedagogical skills and an understanding of the capacity of each child, in order to take learners' thinking forward. Within a dialogic approach, as described by Alexander (2006), learner participation is of the utmost importance, and this in turn addresses motivation and deeper learning, countering the 'recipe threat' mentioned earlier. However, in our opinion, this dialogic approach does not stress adequately the authoritative aspect of science education. The gap between pupils' pre-existing views and the scientific view is often too great to be addressed using the dialogic aspect alone. It is this dimension that is addressed by the communicative approach.

1.1 Communicative Approach and Dialogic Inquiry-Based Teaching

Mortimer and Scott's communicative framework accommodates both dialogic and authoritative approaches in the science classroom. According to Mortimer and Scott (2003), classroom discourse consists of four categories generated from the combination of two dimensions: interactive/noninteractive and authoritative/dialogic (Mortimer & Scott, 2003; Table 1). Whereas the everyday understanding or prior knowledge of learners is often addressed through dialogic approach, the view of science is conveyed through authoritative approach. The interactive/noninteractive dimensions indicate the different ways in which teachers can use talk, whether through reciprocal discussions, question/answer sessions or teacher talk. Scott and Ametller (2007) stress that meaningful science teaching should include both dialogic and authoritative aspects and that the relationship between these two aspects is highly significant. For instance, if discussions are 'opened up' through a dialogic approach and pupils are given the opportunity to work with different ideas, at some point discussions should also be 'closed down' using an authoritative approach in the end when making scientific conclusions. The 'closing down' phase is potentially very important, for example, when clarifying the differences between pupil

Table 1 Communicative approaches and examples

Communicative approach	Example
Authoritative and noninteractive	Teacher: Now, here *we* see a model of a hydrogen gas molecule. The hydrogen gas molecule has two hydrogen atoms with a covalent bond formed by a pair of electrons…
Authoritative and interactive	Teacher: What does the capital letter *E* stand for in this equation?
	Student: For power?
	Teacher: *No! It stands for energy!* Well what about *U*? (Continues with IRE structure)
Dialogic and interactive	Teacher: Can *you* explain what causes the container to explode under pressure? (*Wait time*)
	Student 1: Isn't it like when you put a plastic bottle that's filled with air into water, or like, under the water, and it squashes because of the pressure?
	Teacher: Do you know why that happens, *can you explain further?*
	Student 2: Well it's because… I can't remember.
	Teacher: *That's all right.* I think *it will become clearer during the lesson.*
Dialogic and noninteractive	Teacher: …The energy converts to another form. As *you* said, for example, to sound, against gravitation and other forms, which all were good notions. But when *we* think in terms of energy 'loss'… Well, some is lost as sound, but it's mostly through air friction as heat…the energy changes form to heat when air particles collide with the ball… (Reviews student responses and shifts towards more scientific explanations.)

everyday views and the scientific view in the end. Accordingly, there are also results revealing that increasing the level of openness itself is not always beneficial for student learning (Jiang & McComas, 2015).

To meet the challenge of implementing inquiry-based (Minner, Levy, & Century, 2010) learning with dialogical aspects and educational goals, we developed an exemplary process model to take these different aspects into account (Table 2). The model could also be considered as a theory-based planning tool for dialogic inquiry-based learning with both dialogic and authoritative modes included in the table. The first column of the table lists the different phases of an inquiry. The middle column indicates how each phase connects with key notions from inquiry-based learning, and the third column lists the communicative approach for establishing the structure of opening up and closing down (Scott & Ametller, 2007). Briefly, in order to stand for dialogic, inquiry should include dialogic opening-up phase followed by practical inquiry phase with focus on supporting student-student interaction. What is notable is that authoritative approach is a part of the model. Yet, saving it to the closing-down phase would meet our criteria for dialogic, as too frequently authoritativeness hindering student bringing their views to classroom discussion (e.g. Lehesvuori, Ramnarain, & Viiri, 2017).

The *opening-up phase* includes examining pupils' preconceptions, and even though preconceptions at this point might be considered misconceptions, pupils should be given an opportunity to express them. Using a problem-based approach,

Table 2 The model for dialogic inquiry-based teaching and learning

	Inquiry-based learning	Communicative approach
Initiation phase	Problem-based approach	Opening-up phase
	Considering pupils' preconceptions	Dialogic and interactive
		Dialogic and noninteractive
Practicing phase	Planning	(Emphasis is on pupil-pupil interaction)
	Making hypotheses	
	Collecting information	
	Executing the inquiry	
Reviewing phase	Comparing results to the scientific view	Closing-down phase
	Creating models	Dialogic and noninteractive
	Argumentation	Authoritative and interactive/ noninteractive
	Reinforcing the scientific view	

the teacher could uncover these (mis)conceptions by employing a dialogic approach and opening up problems requiring inquiry. At a later stage, the views can be reflected upon again, reconsidered in the light of the executed inquiry.

The *practicing phase* includes planning, executing and reflecting on the results. Hypotheses are made and tested, and results are discussed among peers. The role of the teacher should be more of a tutor than director, in this way creating the ground for meaningful planning and inquiries. Although pupils are expected to do the thinking, the teacher could still raise questions that guide pupil work and thinking further. In this phase the teacher should focus on encouraging pupil-pupil interaction, and despite important, it is not the focus of this chapter. Scaffolding strategies are reported elsewhere (Lehtinen, Lehesvuori, & Viiri, 2017).

Although in *reviewing phase* more authoritative communication is emphasized, any pre- and misconceptions should be reviewed against the scientific results and theories in order to make the connections between views (e.g. everyday views and the science view) explicit and also highlight possible weaknesses in previous thinking. While different ideas are still being considered, the dialogic approach should also be used. The authoritative approach should continue to be implemented when making the final conclusions about the content and also about the procedure itself. Therefore, when problems are opened up (dialogic approach), they should be also closed down (authoritative approach).

2 Designing and Developing a Training Module for Dialogic IBST

In this section we present how the dialogic approach was integrated to IBST in a science education course for primary school student teachers. The course covers an academic year, beginning in September and ending in March. It includes subject lectures, pedagogical seminars, project work and teaching practice in schools. The course *lectures* contain various elements of science teaching to familiarize primary student teachers with dialogic inquiry-based science teaching and learning such as models in science, pupils' conceptions, evaluation and planning of teaching. These general topics deal with different areas of science (physics, chemistry, biology, geography). Before each lecture, students study the material, familiarizing themselves with the content of the lecture. At the beginning of every lecture, *content tests* evaluate students' prior content knowledge of the given issues. The aim of the content tests is to both activate student prior knowledge and to provide lecturers with information regarding student teacher conceptions.

The pedagogical seminars focus on dialogic inquiry-based learning. The first seminar introduces the course principles, the study project ideas and inquiry-based learning and is then followed by four specifically science education seminars. The science education seminars focussed on the socially and environmentally important topic of climate change. It was hoped that as primary school student teachers are not science specialists, the relevant contextual framework might support student teacher interest in the science concepts and ideas. Furthermore, this topic supports the linking together of physical, geophysical, chemical, biological and environmental aspects of climate change.

The seminars also introduced the students to the following issues: what is science, how science works and how dialogic inquiry-based teaching can be incorporated into science teaching. Moreover, during one of the seminars, student teachers were introduced to the classification of teacher talk based on the communicative approach of Mortimer and Scott (2003). The student teachers were provided with different examples of teacher talk with directions for classifying classroom interaction. They then visited a local primary school and observed a class they were to teach during the implementation phase of the planned teaching sequence. They completed a specific observation form during the science lesson and classified the teacher talk according to the different communicative approaches. The aim behind this classroom-based experience was to prepare the student teachers for the planning and inclusion of different communicative approaches within their own teaching.

The *study project* is the core of this science education course, and its objective is for student teachers to develop a teaching-learning sequence on a single science topic. The project includes science content analysis; discovering pupil ideas regarding the chosen topic; finding, selecting or creating the most appropriate presentations and teaching strategies; and making a plan for a teaching-learning sequence covering several lessons. The study project also requires students to prepare a written report and oral presentation of the main issues of the project. It is hoped that through the practical implementation of the study project, the students draw on and develop their theoretical understanding and beliefs around dialogic inquiry-based science as well as their practical and pedagogical understanding (Clarke & Hollingsworth, 2002).

2.1 Participants

The course outlined above included approximately 120 participants that were divided into six seminar groups of approximately 20 students. Each group had a specific university lecturer to mentor student work. The groups were further divided into subgroups consisting of 4–5 students. The subgroups planned and implemented the teaching practice lesson together. In the following section, the lessons of two subgroups are presented along with the analysis of the lessons based on the dialogic IBST model.

2.2 Recorded Lessons

The two subgroups both were required to select a topic and to plan and implement a science lesson according to the principles of dialogic IBST. The first group worked with the topic of melting polar ice (Class A) and the second group focused on climate change and the life cycle of porridge (Class B). The student teachers' video-recorded lessons were mapped using the interactional graphic providing a visual representation of the lessons with the interactional patterns clearly identified. Supplementary, contextual data included student lesson plans and interviews after the lessons. The lesson plans were then analysed by the second author to see whether and how the students had implemented the ideas of dialogic inquiry-based teaching in their lesson plans (an example plan is provided at the end of the chapter). Following the lessons, stimulated recall interviews were conducted by the second author to gather student perceptions of their aims and successes in their teaching. Once the collected data were mapped with the interactional graphic for each group, it was possible to compare the degree to which the student teachers implemented the principles and practices of IBST.

3 An Interactional Graphic Tool as a Method for Analysing and Reflecting on a Teaching Sequence

This section includes interactional graphs and contextualizing comments from the two student subgroups and a description of Class A's and B's lessons. It is hoped that the interactional graphic examples applied to the context of student teacher science lessons illustrate the greater potential of this tool in studying science-based classroom interaction.

In the interactional graphic tool, classroom communication is characterized by the different communication approaches used by the teacher (see Figs. 1 and 2). The duration of a particular approach can be estimated from the horizontal axis. This opening-up phase leads into periods of practical activities punctuated by further guidelines given by the teacher, in this instance through the reading of a story (Lehesvuori et al., 2013; Ratinen, Viiri, & Lehesvuori, 2013; Ratinen, Viiri, Lehesvuori, & Kokkonen, 2015). The communicative approaches adopted by the teacher towards the end of the lesson indicate the closing-down phase with increased emphasis on the scientific view. Mapping the interactional patterns of the lesson in this way provides an overall picture of classroom talk. The interactional graphics of different lessons can be placed alongside each other supporting comparisons between lessons. Though the interactional graphic does not 'judge' the interactional choices of the teacher, implicit in the representational style of the graphic is the need to reflect the structure of the model (Table 2): the opening-up phase of the inquiry, the practice part of inquiry and closing down the inquiry. The pattern could be applied across an entire course of study on a particular topic, but the case study group of student teachers prepared a 'stand-alone' double lesson during which they employed the IBST approach.

The interactional graphic tool (Figs. 1 and 2) aims to present the different communicative approaches and periods of inquiry within a lesson, in a readily accessible format. The vertical axis of the diagram displays the four different classes of the communicative approach (Mortimer & Scott, 2003) with the fifth, uppermost section of the axis representing the practice phase of the lesson during which pupil-pupil talk is emphasized.

3.1   Class A

The topic of this lesson was the melting of polar ice. The communicative approaches and active inquiry phases of the lesson are graphically presented in Fig. 1. First the four student teachers briefly introduced themselves to the pupils. Then one student teacher (ST1) asked the pupils about the melting of clean and dirty ice (minutes 3–5).

1. ST1: What do you think, which one of the ice blocks melts down faster, dirty or clean?
2. Pupil1: The dirty one?
3. ST1: And why do you think the dirty ice melts down faster?
4. Pupil1: Well, I don't know. It gets warmer; I guess it warms up faster.
5. ST1: Yeah (Rising intonation)
6. Pupil2: Clean one!
7. ST!: Well why do you think the clean one?
8. Pupil2: Well it's cleaner and the molecules somehow warm up faster or something....
9. ST1: Yeah, ok (Rising intonation, waiting for further ideas) (Discussion continues with collecting ideas).

Student teacher collected pupil ideas without directly evaluating the answers, and in this way pupils' prior ideas were taken into account before group work began. This is indicated by teacher prompting questions (turns 3 and 7) and the rising intonation in feedback (turns 5 and 9) standing for invitation for further ideas (Lehesvuori et al., 2017). According to the lesson plan (see procedure 2 in the lesson plan in the Appendix), the communicative approach planned was interactive/dialogic and was also implemented as one.

After brief instructions, pupils started to work in groups on three tasks: the greenhouse effect, a drawing assignment and planning an advertisement. The melting of two different types of ice was measured by the teacher at approximately 15 min intervals, while the whole class was encouraged to make observations.

The second task (minutes 29–44) began with a story about a polar bear and the melting of a north polar glacier. After the story pupils were asked to draw a cartoon based on the story and to think about the consequences of climate change. The third task (minutes 44–68) required that pupils planned an advertisement to encourage the slowing down of climate change. Once again the task was momentarily interrupted for whole class measurements. After this the teaching, sequence continued with a role-play activity (minutes 70–82). Each pupil was given a role (e.g. polar bear, atmosphere, sun-ray, etc.), and every time this character was mentioned in the story, the pupil(s) demonstrated the actions of this character. The pupils seemed very enthusiastic when playing their roles and explaining the reasons for the melting of a polar glacier. In this way, pupils were introduced to the conclusions drawn during the final parts of the lesson.

The final measurements were conducted and observations were made on the quantity of melted ice. During this phase, interactive/dialogic communicative approaches were used, although authoritative passages gently guided the discussions towards conclusions (see turns 3 and 4). Pupils' preconceptions were also addressed (e.g. see turn 7) within the final conclusions (minutes 82–90).

1. ST1: Well which one melted faster?
2. Pupil1: The clean one! Because it… there is more of clean water!
3. ST2: Well, what can you see here (points out collected measurements and reviews some values between clean and dirty ice)…so 35 ml of dirty ice and 25 ml of clean ice has melted during this time.
4. ST3: So which one do you think melts faster as time goes?
5. Pupil2: The dirty one melts faster as the time goes!
6. ST3: Yes (lowing intonation signalling authoritative evaluation)
7. ST1: So, if we would have more time, we could continue the measurements… (continues discussion on the experiment and directs the discussion towards polar ice)… as you brought up in the beginning, the dirty ice warms up faster, thus melting down faster, yet why do you think this happens? Think about clean polar ice, why doesn't it warm up?
8. Pupil3: Well it reflects the rays.
9. Yeah, there is reflection….

The planned procedures, including communicative approaches, can be clearly identified in the teaching sequence, which illustrates the purposeful use of different discursive strategies. During the dialogic episodes, pupil contributions were especially taken into account with a supportive or neutral tone, thus fostering an open climate for further contributions from pupils. In addition to purposefulness, other features of dialogic teaching were also present in both the lesson and lesson plan. For instance, acknowledging pupils' prior ideas and addressing these at the end engage both supportive and cumulative teaching. Furthermore, pupil group work with student teachers acting as co-inquirers aimed to embrace collective and reciprocal approaches to pupil inquiries. However, when this aspect was discussed in the interviews, student teachers identified it as being challenging. They thought that they were renegotiating rather than guiding pupils in a certain direction, indicating the challenge of balancing dialogic and authoritative approaches. This challenge is also evident in the lesson plan. Procedure 4 (see Appendix) indicates a clear conflict between adopting an interactive/dialogic approach and "finding the right answers with teacher's support". However, in conclusion the structure of the planned teaching sequence closely followed the approach for planning and implementing dialogic inquiry-based learning.

3.2   Class B

This class focused on climate change and the life cycle of porridge and its environmental impacts. In the lesson energy consumption of porridge cooking was measured and further discussed its influence to the life cycle of porridge further discussed.

To begin Class B (Fig. 2), the teacher clarified some practical issues and collected pupils' preconceptions about climate change (0–5 min). The class consisted of student teacher-led sessions with the children (5–7 min, 7–15 min, 15–23 min, 23–26 min, 32–36 min and 66–75 min), completion of a worksheet (26–32 min) and the initiation and implementation of student-teacher-orchestrated experiment, cooking porridge in a saucepan with and without a cover (36–38 min, 41–56 min, 56–66 min).

This teaching sequence involving the life cycle of porridge did not appropriately follow the inquiry-based learning approach (Fig. 2). The lesson did not demonstrate the full range of communicative options, and, as can be seen from the communication graphic, practicing phases were completely absent from this lesson. Whereas Class A effectively illustrated the three-part pattern of the model (opening up, inquiry, closing down), Class B had no opening up, maintaining authoritative communication throughout the lesson and omitting any authentic phases of student inquiry or dialogue. Instead of closing down, the discussion in the end was unrelated to the experiment of cooking porridge. In particular, porridge cooking as an example experiment failed because, among other things, the student teachers neglected to measure electricity consumption after cooking. In short, this class represents student teachers' limited understanding of inquiry-based science teaching. The form of interaction was dominantly the kind of presented in latter, authoritative, classroom example of class A. Despite their plans to use dialogic communication for gathering pupils' ideas, dialogic approach was not taking place in the beginning. As a result, closing down was not present in terms of taking these preliminary ideas into account when summing up the lesson.

4   The Use of Interactional Graphic as a Tool for Teacher Professional Development

As important as it is to detect actions, such as questioning and feedback, it would be as important to have a more holistic and dynamic view of the executed lessons, which is an issue the interactional graphic tool is addressing (Lehesvuori et al., 2013). The interactional graphic tool can be used with pre- and in-service teachers to provide a holistic view of the overall procedure from planning to implementing and reflecting on IBST.

Planning Teaching Sequences Evidence suggests that especially novice teachers focus on content in their planning and neglect the communicational aspect, but when introduced to a framework including communicative approaches, they are able to vary their communication in action (Lehesvuori, Viiri, & Rasku-Puttonen, 2011). In this regard, the IBST framework should be particularly useful as a planning tool helping teachers direct their attention to the use of the communicative approach and patterns of discourse in different parts of the lesson and how they create links between different phases of the inquiry.

Video Recording Lessons The technical requirements needed to capture the essentials of teacher-orchestrated communications are not unattainable for regular teachers and schools. A camera located at the back of the classroom combined with a portable teacher microphone has proven sufficient. Another possibility is peer videoing, which would give teachers' experience about being empirical researchers of their own professional development, which may provide a catalyst for reflection and critical dialogue among colleagues (Hartford & MacRuairc, 2008, p. 1890).

Reflection After the implementation of the lessons, a teacher may self-reflect on lessons by following the guidelines of the stimulated recall (interview) technique (STRI technique) (O'Brien, 1993). Another option, however, is to use the IBST graphical tool to visualize the communicative dynamics of the lesson and to see whether the structure of opening up/closing down on the scale required in overall inquiry process was employed. Interpretations can then be supported/challenged/discussed with a colleague or a mentoring teacher who also has access to the lesson videos.

5   Discussion and Conclusion

The IBST graphical tool is a novel means for presenting and reflecting on teaching sequences in pre- and in-service education with potential. As we have demonstrated in the formal analysis of classroom interaction, the graph could visualize the model. The interactional graphics would be useful in pre- and in-service teacher reflections of classroom practice. As illustrated as part of the science education course outlined in this chapter, the student teachers observed a science lesson and made individual notes according to a prepared format. An extension to this task could be for teachers to construct an interactional graphic of the observed lesson. This would encourage teachers to truly engage with the different interactional options and to clarify what these mean in inquiry-based approaches. This should further support teachers in lesson planning and the realization of IBST.

The IBST approach aims to reform teachers' traditionally authoritative view of science teaching, but there are still major challenges ahead when thinking about adopting the dialogic aspect. Aside from the question of time and discipline (Scott, Mortimer, & Aguiar, 2006), the science classroom culture may not be open to dialogic innovations. In order to challenge this prevailing culture, dialogic issues need to be emphasized in initial and in-service teacher education. Often, however, teachers are not able to effectively use the appropriate pedagogical strategies discussed in teacher professional development courses. Indeed, teachers' perceptions and methods of teaching are deeply grounded in their own experiences of school as pupils (Abell, 2007). If teacher professional development does not explicitly address different approaches to teaching, there is a danger that those beliefs will persist throughout teacher education and teaching service (Fajet, Bello, Leftwich, Mesler, & Shaver, 2005). On this basis, increasing teacher awareness and opportunities to engage with and reflect on practice at both pre-service and in-service levels is essential in initiating reform of practice (Clarke & Hollingsworth, 2002). The theoretical model and interactional graphic are presented here as two key tools in the introduction and development of dialogic innovations in science education applicable to both initial and in-service teachers.

Appendix: Lesson Plan

Level: 6	Time and date:	Topic/objectives (science): Climate change and its influence on the life of glaciers		
	2×45min (90 min) 22.2.2011	**Summary of special education** Every pupil can participate according to her/his abilities. Group work: attention to individuals		
Educational and learning objectives	**Learning process, content, time management, specialization**	**Procedure**		**Evaluation and feedback**

(continued)

Educational objectives: Dialogue and interaction Group work skills Stimulation of individual thinking Learning objectives: Understand the causal relation of climate change Understand the greenhouse effect vs. climate change Understand the complexity of climate change Accessing the experiment: ice cube demonstration Small group work supports the consideration of individuals and their needs	1. Opening class: topic presentation (5 min) 2. Setting experiment (10 min) Experimental design Linking to glaciers Making hypotheses Tabling on blackboard 3. Forming groups (each five persons) (2 min) 4. First task: What causes climate change? Completing picture with teacher guidance (15 min)* 5. Second task: What follows climate change? Drawing picture based on story (15 min)* 6. Third task: How to prevent climate change? Making an ad based on given material (15 min)* 7. Synthesis: What is it? Pupils play a drama (15 min) 8. Reviewing the experiment: Is the hypothesis true? Why or why not? (13 min) *) At the end of each task, the melting ice is observed	1. Noninteractive/ authoritative Teachers present the topic of class 2. Noninteractive/ authoritative Setting and explaining experiment Dialogic: collecting and discussing of hypothesis 3. Noninteractive/ authoritative 4. Interactive/dialogic Group discussion with teacher tutoring Finding the right answers with teacher support 5. Noninteractive/ authoritative Teacher reads the story to pupils 6. Interactive/dialogic Pupils negotiate the story and draw a picture 7. Interactive/dialogic Pupils seek information and teacher help if needed 8. Noninteractive/ authoritative Teacher reads the story to the end 9. Interactive: pupils act the drama 10. Dialogic Pupils explain their observation of ice melting Discussion about hypothesis	Teachers support and encourage during every task- > direct feedback E.g. understanding the meaning of the experiment- > Why we did as we did? Evaluation new conceptualization (new drawings): Do pupils understand the connection and difference between the greenhouse effect and climate change? Has pupil knowledge of climate change increased? Do pupils understand the connection between climate change and glaciers melting? Include both individual and group evaluation plus feedback

References

Abell, S. (2007). Research on science teacher knowledge. In S. Abell & N. Lederman (Eds.), *Handbook of research on science education* (pp. 1105–1149). Mahwah, NJ: Lawrence Erlbaum Associates.

Alexander, R. (2006). *Towards dialogic teaching* (3rd ed.). New York: Dialogos.

Clarke, D., & Hollingsworth, H. (2002). Elaborating a model of teacher professional growth. *Teaching and Teacher Education, 18*(8), 947–967.

Fajet, W., Bello, M., Leftwich, S. A., Mesler, J. L., & Shaver, A. N. (2005). Pre-service teachers' perceptions in beginning education classes. *Teaching and Teacher Education, 21*, 717–727.

Hartford, J., & MacRuairc, G. (2008). Engaging student teachers in meaningful reflective practice. *Teaching and Teacher Education, 24*, 1884–1892.

Jiang, F., & McComas, W. F. (2015). The effects of inquiry teaching on student science achievement and attitudes: Evidence from propensity score analysis of PISA data. *International Journal of Science Education, 37*(3), 554–576.

Lehesvuori, S., Ramnarain, U., & Viiri, J. (2017). Challenging transmission modes of teaching in science classrooms: Enhancing learner-centredness through dialogicity. *Research in Science Education*, 1–21. https://doi.org/10.1007/s11165-016-9598-7.

Lehesvuori, S., Viiri, J., & Rasku-Puttonen, H. (2011). Introducing dialogic teaching to science student teachers. *Journal of Science Teacher Education, 22*(8), 705–727.

Lehesvuori, S., Viiri, J., Rasku-Puttonen, H., Moate, J., & Helaakoski, J. (2013). Visualizing communication structures in science classrooms: Tracing cumulativity in teacher-led whole class discussions. *Journal of Research in Science Teaching, 50*(8), 912–939.

Lehtinen, A., Lehesvuori, S., & Viiri, J. (2017). The connection between forms of guidance for inquiry-based learning and the communicative approaches applied – A case study in the context of pre-service teachers. *Research in Science Education*, 1–21. https://doi.org/10.1007/s11165-017-9666-7.

Minner, D. D., Levy, A. J., & Century, J. (2010). Inquiry-based science instruction – What is it and does it matter? Results from research synthesis from years 1984 to 2002. *Journal of Research in Science Teaching, 47*(4), 474–496.

Mortimer, E. F., & Scott, P. (2003). *Meaning making in science classrooms*. Milton Keynes, UK: Open University Press.

National Research Council. (2000). *National science education standards*. Washington, DC: National Academy Press.

O'Brien, J. (1993). Action research through stimulated recall. *Research in Science Education, 23*, 214–221.

Oliveira, A. W. (2009). Developing elementary teachers' understandings of hedges and Personal pronouns in inquiry-based science classroom discourse. *Journal of Research in Science Education, 8*(2), 247–269.

Ratinen, I., Viiri, J., & Lehesvuori, S. (2013). Primary school student Teachers' understanding of climate change: Comparing the results given by concept maps and communication analysis. *Research in Science Education, 43*(5), 1801–1823.

Ratinen, I., Viiri, J., Lehesvuori, S., & Kokkonen, T. (2015). Primary student-teachers' practical knowledge of inquiry-based science teaching and classroom communication of climate change. *International Journal of Environmental and Science Education, 10*(4), 561–582.

Sadeh, I., & Zion, M. (2009). The development of dynamic inquiry performances within an open inquiry setting: A comparison to guided inquiry setting. *Journal of Research in Science Teaching, 40*(10), 1137–1116.

Scott, P., & Ametller, J. (2007). Teaching science in a meaningful way: Striking a balance between 'opening up' and 'closing down' classroom talk. *School Science Review, 88*(324), 77–83.

Scott, P. H., Mortimer, E. F., & Aguiar, O. G. (2006). The tension between authoritative and dialogic discourse: A fundamental characteristic of meaning making interactions in high school science lessons. *Science Education, 90*(4), 605–631.

Part III
Fostering Teachers' Competences in Cross-Domain Scientific Inquiry

Promoting Pre-service Teachers' Ideas About Nature of Science Through Science-Related Media Reports

Gultekin Cakmakci and Yalcin Yalaki

1 Introduction

Policy and reform documents about science education in the last two decades have focused on fostering scientific literacy among all students and also promoted inquiry-based science teaching and learning (IBST/L) as a recommended way of teaching science to achieve scientific literacy (Turkish Ministry of National Education (MEB) 2018; Millar, 2006; National Research Council (NRC), 2000). Today, it is widely accepted that science education should equip students with the knowledge and skills to become scientifically literate citizens (Elliott, 2006). Although the meaning of scientific literacy in the context of school science has been debated, it is widely agreed that understanding nature of science (NOS) is an essential aspect of public engagement with science and scientific literacy (Driver, Leach, Millar, & Scott, 1996). NOS refers to "the epistemology and sociology of science, science as a way of knowing, or the values and beliefs inherent to scientific knowledge and its development" (Lederman, Abd-El-Khalick, Bell, & Schwartz, 2002, p. 498). Because of its significance, the Next Generation Science Standards (NGSS) developed by the lead states in the USA (2013) incorporated NOS tenets in the standards. Some of the tenets were considered as part of the core practices, while others were considered as part of the crosscutting concepts, which are two of the main dimensions of the standards. Even though IBST/L has been shown to be an effective way of teaching science (Minner, Levy, & Century, 2010), it does not necessarily help students to understand NOS or achieve scientific literacy (Lederman, 1999). As cited in the NGSS (2013), Conant provided the same argument in 1951: "…Being well informed about science is not the same thing as understanding science, though the two propositions are not antithetical…" (Conant, 1951, p. 4). As

G. Cakmakci (✉) · Y. Yalaki
Faculty of Education, Hacettepe University, Ankara, Turkey
e-mail: cakmakci@hacettepe.edu.tr; yyalaki@hacettepe.edu.tr

© Springer International Publishing AG, part of Springer Nature 2018
O. E. Tsivitanidou et al. (eds.), *Professional Development for Inquiry-Based Science Teaching and Learning*, Contributions from Science Education Research 5, https://doi.org/10.1007/978-3-319-91406-0_8

discussed in the first chapter of this book, one of the opportunities IBST/L provides is the teaching and learning of NOS; however, this should often be supported by explicit pedagogical approaches (Khishfe & Abd-El-Khalick, 2002). This chapter discusses an effective approach to teaching and learning of NOS. The methods suggested in this chapter utilize science-related media reports as a mediating artefact, or tool, for promoting pre-service teachers' (PSTs) conceptions of science and NOS as well as formative assessment as a support mechanism for learning. Exemplary cases are presented to illustrate how a particular media report can be used in practice for teaching NOS.

1.1 Promoting Inquiry in Science

IBST/L is being promoted as a better way of teaching science in many countries including EU members, the USA, Australia, Turkey and others. European Commission (EC) has financed several high-budget projects that specifically focus on improving and promoting IBST/L. The interest in IBST/L comes from a desire to make science education more relevant, meaningful and motivating for students in all grade levels. The term, inquiry, has been used prominently in science education, and it refers to the activities of students and teachers that involve a science-related investigation (Minner et al., 2010; Pedaste et al., 2015). There are different ways of utilizing inquiry in science education, but in any case there are essential features of inquiry in the classroom, which are described by the National Research Council (NRC):

- Learners are engaged by scientifically oriented questions.
- Learners give priority to evidence, which allows them to develop and evaluate explanations that address scientifically oriented questions.
- Learners formulate explanations from evidence to address scientifically oriented questions.
- Learners evaluate their explanations in light of alternative explanations, particularly those reflecting scientific understanding.
- Learners communicate and justify their proposed explanations (NRC, 2000, p. 25).

More recently, NGSS (2013) described inquiry practices as "science and engineering practices" which should be developed throughout K–12 science education. The practices are listed as:

- Asking questions (for science) and defining problems (for engineering)
- Developing and using models
- Planning and carrying out investigations
- Analysing and interpreting data
- Using mathematics and computational thinking

- Constructing explanations (for science) and designing solutions (for engineering)
- Engaging in argument from evidence
- Obtaining, evaluating and communicating information (NGSS, 2013: Appendix F, p. 1)

The above features or practices are essential when utilizing inquiry in science education, and they can be used to differentiate inquiry activities from other hands-on classroom activities. The utilization of inquiry provides opportunities for discussion of various aspects of the scientific process (i.e. aspects of NOS) with students. For example, when different groups of students approach an inquiry question differently or come up with different explanations, teachers may discuss the subjectivity involved in scientific inquiry or the role of imagination and creativity. However, without explicitly stating these aspects, students may not realize them during the inquiry process. Therefore, the focus of this chapter is not to explain how to utilize inquiry in the classroom but to focus on the teaching of NOS. IBST/L and NOS are concepts that provide a more complete understanding of science when utilized together in the science classroom (Lederman, 2006). NOS provides a philosophical background for scientific inquiry, and it is considered an integral part of today's scientific literacy and science curriculum. Although there is no single and universally accepted definition of NOS, according to Lederman (2007) and McComas (2017), significant academic consensus has been achieved on the aspects of NOS to be taught in school science. These aspects state that the scientific knowledge is:

- (NOS-1) Both reliable (one can have confidence in scientific knowledge) and tentative (subject to change in light of new evidence or reconceptualization of present evidence).
- (NOS-2) Empirically based (based on and/or derived from observations of the natural world).
- (NOS-3) Not based on a single, universal scientific method that captures the complexity and diversity of scientific investigations.
- (NOS-4) Subjective and/or theory-laden (scientists' values, knowledge and prior experience as well as contemporary scientific perspectives influence their observations and the collection and interpretation of empirical data).
- (NOS-5) Partly the product of human inference, imagination and creativity (involves invention of explanations).
- (NOS-6) Socially and culturally embedded.
- (NOS-7) Based on a distinction between *observations* and *inferences*.
- (NOS-8) Subject to distinctions between the functions of, and relationships between, scientific *theories* and scientific *laws* (Abd-El-Khalick & Lederman, 2000, p. 1063).
- (NOS-9) Characterized by proper scientific explanations or arguments that involve the coordination of the evidence/data and the claim to support or refute an explanatory conclusion, model or prediction (Osborne, Erduran, & Simon,

2004). In other words, it involves reasoning to produce new knowledge about the natural world.

Of course there are other conceptions of NOS that include more aspects of science or provide a more complex vision of science. However, the discussions of these different conceptions of NOS are beyond the scope of this chapter. The so-called consensus view as stated above is considered for the purposes of this chapter.

An example case of a classroom application of inquiry, provided by NRC, is briefly summarized below to exemplify how NOS tenets are relevant in inquiry activities:

> Mrs. Graham's fifth-grade students express curiosity towards the appearances of three trees in the school yard. Even though these trees were side by side on a hill and they were same type of trees, one of them shed all of its leaves, one of them had lost some of its leaves and had multi coloured leaves while the third had lush green leaves. Students wanted to know what was wrong with the trees. Mrs. Graham took her students' curiosity as an opportunity to start an investigation. She grouped her students and asked them to come up with possible explanations. Every group proposed an explanation such as, it has something to do with sunlight; it must be too much water, too little water, it may be insects, it can be the age of the trees etc. Mrs. Graham asked her students to decide which of these ideas were suitable for an investigation. After selecting ideas to be investigated, she asked her students to pick one idea and work in groups to plan and conduct a simple investigation to see if they could find evidence to support their ideas. In three weeks, students carried out their investigations, came up with explanations based on evidence, discussed in the classroom to consider other explanations, decided which explanation was mostly supported by evidence and finally they tested their explanation. They found that the trees were watered inappropriately. (NRC, 2000, p. 6–11)

This vignette is an example of an open inquiry applied in a fifth-grade classroom. When Mrs. Graham's vignette is considered, some of the NOS aspects can be seen in her classroom inquiry activities. For example, *NOS-2* in the above list indicates that scientific knowledge is empirically based. Mrs. Graham's students utilize this aspect of scientific knowledge by constructing evidence-based explanations. It can be argued that each group in Mrs. Graham's classroom approaches the planning and conducting phase of their investigation in a different way, which could be linked to NOS-3, indicating that scientific investigations may be very different from one another and there is no universal scientific method. Students came up with different ideas and explanations regarding the same evidence and discussed and communicated their ideas in the classroom. This could be linked to *NOS-4*, indicating that scientific knowledge can be subjective, and different scientists may interpret the same data differently. Students' imagination and creativity play an important role in planning and conducting their investigations, which could be linked to *NOS-5*, where scientific knowledge is partly the product of human inference, imagination and creativity. Finally, in Mrs. Graham's class, students made observations and came up with inferences, which provided a good opportunity to teach them the difference between inferences and observations (*NOS-7*). Many parallels and relations can therefore be drawn between IBST/L and teaching of NOS. As seen in this example, understanding of NOS provides the necessary foundations for inquiry, and an inquiry environment in a classroom is conducive to teaching and learning of NOS. Therefore, science teachers' NOS understanding and their skills to teach it should complement their skills for applying inquiry in their classrooms.

1.2   Popular Media as an Instructional Tool for Teaching Science and Its Nature

Understanding of NOS is an important part of scientific literacy, and it is also essential in IBST/L (Minner et al., 2010). According to Linn, Davis, and Bell (2004), inquiry is "the intentional process of diagnosing problems, critiquing experiments, distinguishing alternatives, planning investigations, researching conjectures, searching for information, constructing models, debating with peers, and forming coherent arguments". NGSS (2013) list of inquiry practices mentioned in the previous section is similar. In both views, inquiry consists of asking questions about nature and finding answers with the methods of science. Using the methods of science requires acceptance of the epistemological and ontological assumptions of science concerning nature, which are rooted in NOS. For example, the following questions are an integral part of scientific activity:

- How is knowledge obtained in science? Is it evidence based? What is the role of observations vs. deductions?
- Is the world independent of the knower? Or is it subjective and theory-laden?
- Do we collect real knowledge from nature? Or is knowledge created by social and cultural symbols?
- What are the ethical considerations of scientific investigation?

When one engages in inquiry, these questions quickly become important matters to consider. Therefore, understanding of NOS is essential in IBST/L, and without an adequate NOS understanding, IBST/L would lack its foundations.

Some researchers have initiated a pioneering effort to enhance the attractiveness and relevance of NOS to students by using science-related news in the classroom (Elliott, 2006; Jarman & McClune, 2007; Norris, Phillips, & Korpan, 2003; Ratcliffe & Grace, 2003; Storksdieck, 2016). The underlying idea behind these studies is that bringing contemporary and cutting-edge science into the classroom can help form a valuable bridge between the real world and the school science (Jarman & McClune, 2007; Seckin Kapucu, Cakmakci, & Aydogdu, 2015). We therefore explored the use of media reports of scientific research as a context to enhance PSTs' conceptions of scientific literacy, in particular, their conceptions of NOS. Our goal was to use media reports as an instructional tool to help PSTs, and, in return, to their future students to become better informed and more discerning consumers of scientific information and to increase their motivation and willingness to learn science and NOS (Ford, 2009; Jarman & McClune, 2007; Ratcliffe & Grace, 2003). We put a particular emphasis on argumentation skills, which included evaluating the credibility of evidence; establishing the validity of explanatory conclusions, models or predictions; and evaluating sources of both conclusive and inconclusive science (Osborne et al., 2004; Sadler, 2006). These are also essential skills in scientific inquiry. Having informed ideas about epistemology of science and advanced argumentation skills is necessary to grasp the underlying ideas behind media reports of science.

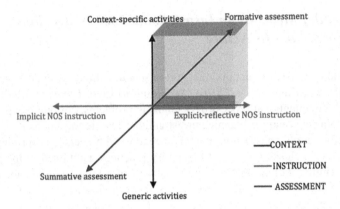

Fig. 1 Implicit vs. explicit-reflective NOS instruction, generic vs. context-specific activities for teaching NOS and summative vs. formative assessment of NOS learning. (Note: Shaded area in the square shows the "main" approaches that were used in this study)

1.3 Theoretical Foundations

Studies showed that people who engage in scientific inquiry alone (even though it implicitly refers to tenets of NOS) do not necessarily develop a contemporary understanding of NOS (Lederman, 1999). Therefore, researchers have usually used an *explicit-and-reflective approach* for developing students' NOS views rather than an *implicit approach* that utilizes hands-on or inquiry science activities lacking explicit references to NOS (Lederman, 2007) (see Fig. 1). In this respect, several researchers agreed that teaching about NOS is important and it should be addressed explicitly and reflectively within *contextualized* activities rather than only within *generic (decontextualized, domain-general)* activities (Cakmakci, 2012; Clough, 2006; Duschl, 2000; Leach, Hind, & Ryder, 2003) (see Fig. 1). They claim that generic activities have limitations when it comes to engaging participants in connecting particular aspects of NOS to science content and science process skills. They also argue that not only should aspects of NOS be explicitly taught, but they should also be explicitly assessed within relevant contexts. In this chapter, we reported our study in which an explicit-reflective NOS instruction was used as a pedagogical framework in the context of media reports about scientific research. Nonetheless, it should be pointed out that there are also alternative views suggesting that NOS understanding is manifested and leveraged through scientific investigations what might be called through a "grasp of science" (Ford, 2009). Ford argues that scientists or students who have a grasp of practice may or may not be able to translate this knowing into an explicit form with a traditional assessment tools; therefore, assessment of NOS within relevant contexts by specifically investigating relationships between construction and critique of explanations is crucial.

Researchers often have "one shot" at data collection (summative assessment) from any individual learners' understanding of NOS (Lederman, 2007) (see Fig. 1). In the literature, most of the research measures students' learning of NOS using summative evaluation, sometimes supported by interviews. In our pre-service teacher education programme, we used open-ended summative assessment instruments such as *Views of Nature of Science Questionnaire* (*VNOS, Version D+*)[1] in conjunction with interviews to probe PSTs' NOS views (Lederman et al., 2002) as well as formative assessment methods as a way to examine and improve PSTs' NOS views (Harrison, 2015). Cowie and Bell (1999, p. 101) define formative assessment as "the process used by teachers and students to recognize and respond to student learning in order to enhance that learning, during the learning". Formative assessment is a process that involves setting quality goals for student learning, providing quality feedback and informing instruction with assessment and reflective practice (Black & Wiliam, 1998; Cowie & Bell, 1999; Nicol & Macfarlane-Dick, 2006). Nicol and Macfarlane-Dick (2006) identified the following principles of good feedback practice in formative assessment:

Good feedback practice:

1. Helps clarify what good performance is (goals, criteria and expected standards)
2. Facilitates the development of self-assessment (reflection) in learning
3. Delivers high-quality information to students about their learning
4. Encourages teacher and peer dialogue around learning
5. Encourages positive motivational beliefs and self-esteem
6. Provides opportunities to close the gap between current and desired performance
7. Provides information to teachers that can be used to help shape teaching (p. 205)

The CERI report published by OECD (2005) also emphasizes the importance of peer feedback in addition to these feedback practices. We utilized these principles in our study and provided the details of the formative assessment procedures we used in the third "Exemplary Case" below. This procedure involved assessing PSTs' NOS views, based on the concrete examples in the science-related media reports, to encourage them to elaborate their ideas (Yalaki & Cakmakci, 2011). Media reports provided us with a context that made it easier to communicate ideas and facilitate the formative process.

Figure 1 summarizes our approach to teaching NOS in classrooms. *Explicit-reflective NOS instruction* with *context-specific activities* and increased use of *formative assessment* have potential to promote understanding of NOS (see the shaded area in Fig. 1) (Cakmakci, 2012; Cakmakci & Yalaki, 2011, 2012; Yalaki & Cakmakci, 2011).

---

[1]VNOS, Version D+ is available at https://science.iit.edu/sites/science/files/elements/mse/pdfs/VNOS-D%2B.pdf

2   Methodology

To put our views about teaching and learning of NOS in practice, in 2 consecutive years (2009 and 2010), we developed resources and strategies to utilize media reports in teaching NOS. We implemented and tried these resources and strategies in the third year of a 4-year-long teacher education programme with 118 prospective elementary school teachers in 4 classrooms in a science method course in the fall semester that continued for 14 weeks. In the first few weeks, at least a class hour (50 min) was used to give the PSTs an explicit-reflective NOS instruction, and then they were introduced to some techniques to select and use media reports of science to illustrate and discuss scientific principles, processes and ideas about science and NOS tenets. They were asked to find science-related media reports that were about a scientific investigation, which involved information about scientific methods, observations, evidence, subjects, inferences, findings and claims from well-known news sources (print media, Internet, TV, radio, etc.). This type of news report proved to be most useful, with sufficient usable material to teach NOS and science concepts. PSTs were discouraged from using news that only provided information about a technological advancement or a scientific concept or phenomenon, since such news reports lacked the details and materials needed. PSTs conducted media report analysis in three stages. The first stage was called *Surface Analysis* which aimed to establish reliability of the news and its source, the second stage was called *Analysis Based on NOS* which aimed to find out which NOS tenets may be discerned in the news report, and finally the third stage was called *Analysis Based on the Science Curriculum* which aimed to connect the science concepts in the news report with the science curriculum. An example of an appropriate article and the method of detailed analysis are provided in Appendix 1.

In addition, the PSTs were asked to critically evaluate a media report in a group, make a presentation to their peers in the classroom and afterwards write a report about their classroom teaching. Each group of PSTs presented their work in one class hour. One of the aims of this activity was to improve PSTs' pedagogical content knowledge in regard to NOS (Yalaki & Cakmakci, 2011). The Views of Nature of Science (VNOS) Questionnaire (Lederman et al., 2002) in conjunction with individual interviews was used to assess PSTs' NOS views at the beginning and end of the course. However, this is not the focus of this chapter, since the data has been presented elsewhere (Cakmakci & Yalaki, 2011; Yalaki & Cakmakci, 2011). The results of these studies revealed that compared to their ideas at the beginning of the course, many PSTs developed informed ideas about NOS throughout the course. Formative assessment procedures were introduced in one of the classes as the third dimension in organizing a NOS course (Fig. 1). Below, two examples of how science-related media reports were used and one example of how formative assessment was utilized in our study are provided. The example provided in Appendix 1 provides a full analysis of a news article, which can be used for supporting the learning of NOS in parallel with science content knowledge, and it describes how a news article can be integrated into the science curriculum.

3 Exemplary Cases

The following examples are provided to support this chapter's aims which are (1) to describe how media reports of science were incorporated into a science methods course, (2) to show how formative assessment was used during the course and (3) to explore the participants' views about NOS. Note that whenever appropriate, the relevant NOS aspects were made explicit to PSTs during instructions before they were engaged in classroom discussions about NOS using media reports as a context.

3.1 Example 1: Using Media News to Teach Socially and Culturally Embedded Nature of Science

In this example, an episode of classroom discussion with PSTs is provided as an example to show how media reports of scientific research were used to teach one of the tenets of NOS. The PSTs, for their first assignment, wrote an analysis report about an online newspaper article titled "The Latest on the Relationship Between Cancer and Cell Phones" taken from a news website (CNN Turk, 2010), a well-known news source in Turkey. PSTs conducted their analysis based on a scheme developed by Cakmakci and Yalaki (2011) (see Appendix 1 as an example). The following discussions took place in the classroom after the instructor read PSTs' analysis of the news article and provided written feedback to them about their analysis as part of a formative assessment process. Note that the instructor aimed to make the target NOS tenet, which was "scientific studies are influenced by their social and cultural environment" (NOS-6) as explicit as possible during the discussion. The excerpt of the news article that guided the discussion is also given below:

> The Interphone Study Group has been conducting epidemiologic research with more than five thousand cell phone users in 13 countries for the last 10 years. The member states in this group are Germany, Denmark, Australia, France, Finland, England, Israel, Italy, Japan, Canada, New Zealand, Norway and Sweden. About 100 scientists from the member states have participated in the Interphone study. (CNN Turk, 2010)

1. *Instructor:* Do we see science affected by the social and cultural environment regarding NOS in this news?
2. *PST 1:* No.
3. *Instructor:* You say no and your friend there also had the same conclusion. Are you sure? If you think about it ... What kind of a social and cultural environment can we talk about here? For example, look at the countries where this study was conducted. What kind of countries are these? In which countries this study was conducted?
4. *PST 1:* England, Japan, Canada...
5. *PST 2:* Germany, France...
6. *PST 1:* Developed, rich countries...
7. *Instructor:* Developed, rich countries... So, would you think about doing this study in a poor and underdeveloped country?
8. *PST 1:* If there is no phone, they can't do it. But it is logical to do this where there is more technology use.

9. *Instructor:* So it is logical to do this research in those countries? Why is it logical? Social and cultural environment in those countries makes this research logical in those places. It wouldn't be logical to do this study in a place where cell phone use is not common.

10. *PST 1:* It would be illogical, yes.

11. *Instructor:* So can we say that scientific studies are influenced by social and cultural environments?

12. *PST 3:* Yes.

13. *PST 2:* In that case we can say that.

14. *Instructor:* In this news, we can make a comment like this: We can say that science is influenced by social and cultural environment like this: the reason why this research is conducted in these countries is that these countries are rich countries where there is intense technology use and cell phone use is common.

The above conversation lines are numbered to make it easier to explain the progression of the discussions. In line 1, the instructor tried to explicitly connect a NOS tenet with the content of the news (the context) by asking PSTs a question about it. PSTs failed to see a connection, and after a PST said "No" (line 2), the instructor gave PSTs clues to help them make a connection between the target NOS tenet and the context. Then PSTs discovered a connection between the social and cultural environment and the scientific investigation in the news (lines 4 through 8) with the help of the instructor. The instructor explicitly expressed the reasoning for this connection (line 9), and PSTs agreed with this reasoning (lines 10 through 13). Finally, the instructor summed up the discussion by making the social and cultural influence on scientific investigation explicit (line 14). This example took place in the beginning of the semester, and the instructor took the lead in the discussion since the PSTs were just introduced to the process.

In a high-quality newspaper article about a scientific investigation, it is possible to see a few of the NOS tenets expressed in various ways, as presented above. In the above example, PSTs were learning to use newspaper articles as a context to capture concrete examples of how NOS tenets may be observed in a real-life situation. In this case, they were able to observe that scientific investigations are influenced by the social and cultural environment. At the same time, they were learning about how to introduce NOS and inquiry in a classroom context.

3.2   Example 2: Using Media News to Teach the Difference Between Observation and Inference

This example shows how a media report titled "Did the First Modern Human Appeared 400 Thousand Years Ago?" was used to discuss the difference between observation and inference with PSTs. The news article was taken from the Cumhuriyet newspaper website (Cumhuriyet, 2010), a well-known newspaper in Turkey. The following excerpt summarizes the content of the news:

Jerusalem- According to a new thesis by the scientists, a 400,000-year-old tooth found in the Qesem Cave, 12 kilometres east of Tel Aviv in Ros Ha'Ayin, bears properties that may change the evolutionary history of humans. A team from Tel Aviv University conducted research in the cave, which is in the middle of Israel, and uncovered a tooth which, they indicated to be 400,000 years old and bears many features that resemble "Homo Sapiens," a scientific term for modern humans. (Cumhuriyet, 2010)

A group of three PSTs analysed this news report based on a scheme developed by Cakmakci and Yalaki (2011) and prepared a presentation for the classroom about their analysis. After all of the PSTs read the news, the following discussion took place in the classroom during the presentation:

1. *PST 1 (presenting):* Is the difference between observation and inference apparent [in the news]?
2. *PST 2:* A tooth was found, which resulted from an observation. But it is not certain if it belonged to a human.
3. *Instructor:* A tooth was found and the properties of this tooth were observed. These are observations. What are the inferences?
4. *PST 2:* [One of the inferences is that] it belongs to a human.
5. *PST 3:* They say that it belongs to a human.
6. *PST 4:* It is 400,000 years old.
7. *Instructor:* The fact that it is 400,000 years old, is this observation or inference?
8. *PST 5:* It is inference.
9. *PST 6:* I think it is observation.
10. *PST 5:* No, it is inference.
11. *Instructor:* How can you infer that it is 400,000 years old?
12. *PST 7:* The calculations in various stages are inferred.
13. *Instructor:* Okay, for dating very old stuff, they look at the radioactive substances. In the environment, there are many atoms with certain properties, as also in human body. For example, there is something called carbon dating. By this, ages of things up to 50,000 years can be determined. By looking at other atoms, even older things can be dated. So the age of 400,000 years can be observed [based on radiometric dating]. However, the inferences that they came up from this are, first of all, it belongs to a human. What else?
14. *PST 8:* The origin of humans could be today's Israel.
15. *Instructor:* Yes, it says that modern humans' origin could be today's Israel. This is also an inference. We don't know whether this is true or not.
16. *PST 9:* First modern humans appeared 400,000 years ago.
17. *Instructor:* Right, first humans appeared 400,000 years ago...
18. *PST 9:* But this is an inference isn't it?
19. *Instructor:* Right, inference.
20. *PST 5:* But... didn't we say this is observation?
21. *Instructor:* The determined age of the tooth is an observation. But by looking at this, saying that modern humans appeared 400,000 years ago is an inference. How can we know that even older evidence will not be found? Or maybe there is a mistake in the observation, may be the tooth is not that old.
22. *PST 10:* It is not even certain that it is a human tooth.
23. *Instructor:* So we can say that "the first modern humans appeared 400,000 years ago" is an inference.

As can be seen from this discussion, the news article provided a very useful context to discuss the difference between observations and inference (NOS-7) in the classroom. In line 15, one of the presenting PSTs asked the classroom if they could see the difference between observation and inference in the news report they read. A PST quickly indicated the observation mentioned in the news and an inference from it without explicitly mentioning what the inference was (line 16). Then the instructor shaped the discussion to make inference more explicit (line 17) after which PSTs expressed the inference explicitly (lines 18 and 19). Another PST's suggestion that the determined age of the tooth is an inference, a discussion with agreements and disagreements took place (lines 20 through 24). At this point, to make the difference between inference and observation clear, the instructor asked a question to the class (line 25). Instructor's content knowledge is very important in mediating a discussion as can be seen in line 27, the instructor used his knowledge of radiometric dating to clear up some points. If this was a chemistry course, at this point the instructor could have introduced concepts such as atomic structure, radioactivity, etc. PSTs continued to find out what other inferences existed in the news article (lines 29 through 33) after which a PST's question showed her confusion (line 34). This gave the instructor more opportunity to explain further the difference between observations and inferences (line 35 and 37).

Note that the instructor tried to be as explicit as possible to explain the difference between the observation and inference while using the media news report as a medium for generating discussions. This classroom episode provides an example of how one of the NOS tenets, in this case the difference between observation and inference, can be taught using media news reports.

3.3   Example 3: Using Formative Assessment as a Reflective Process to Help Learning of NOS

In our study, besides using media news reports as a context to discuss about NOS, formative assessment was utilized to give feedback to PSTs about their learning in one of the science methods classes. Three formative assessment strategies were used to enhance PSTs' learning after they received instruction about NOS and how to analyse and use media news reports as a context for teaching NOS. In the first strategy, PSTs were given a newspaper report and asked to analyse it in terms of its quality and the NOS tenets that could be observed. Instructor gave written and oral feedback to PSTs about their analysis, which provided them suggestions for how to improve their analysis. In the second strategy, PSTs were given another media news report and asked to analyse it. This time, they used the experience they gained during the first application of formative assessment to read and give feedback to their peers' analysis reports. The instructor then read the feedback given to peers and provided his feedback for the process in a classroom discussion. In the third strategy, PSTs were given another media report to analyse. This time a classroom discussion took place about the news, after which they were asked to self-assess their first analysis and make changes if they needed.

In brief, the formative assessment strategies that were employed in our class-room applications included instructor feedback, peer feedback and self-assessment. The following examples are provided from the first application of formative feed-back to show what PSTs have written in their analysis and what feedback the instructor provided. In this process, PSTs analysed the media news report men-tioned in the first exemplary case above, which was titled "The Latest on the Relationship Between Cancer and Cell Phones". The following excerpt was taken from the news article:

> [Dr. Christopher Wild, director of International Agency for Research on Cancer, explained, "An increased risk of brain cancer has not been established using the data from Interphone. However, observations at the highest level of cumulative call time and the changing patterns of mobile phone use since the period studied by Interphone, particularly in young people, mean that further investigation of mobile phone use and brain cancer risk is merited". (CNN Turk, 2010)]

A quote from a PST's analysis of this news report was:

> In my opinion, the news itself is not clear and because of this I don't think the news is very reliable. At first it is said that there is no risk [in using cell phones] but then it is said that further research is needed regarding different usage conditions, different usage habits and other factors.

From the quote above, it seemed the PST assessed the reliability of a news report by the certainty in the results of the scientific investigation. The instructor wrote the following question as feedback regarding the quote above:

> Does the fact that no certain result is given in the news mean that the news itself is not reli-able? Do the results of a scientific investigation have to be certain?

The purpose of the instructor here was to help the PST to critically think about the nature of scientific investigation and scientific findings and also the difference between the reliability of a news report versus the reliability of scientific investiga-tion. However, the feedback itself was not enough to support learning if it was not followed by action by the instructor and PSTs to change and improve learning. In this case, the feedback was followed by classroom discussions about the PSTs' analysis and the feedback given, which was followed by new media report analysis activities. PSTs were asked to improve their next analysis based on the feedback they received.

Another quote from another PST's analysis of the same news report was:

> This news shows a revolutionary change, which shows the tentative nature of science.

To which the instructor wrote the following feedback:

> Is this a revolutionary change? What is revolutionary change?

It seemed like the PST did not understand the difference between revolutionary and evolutionary change in science, and thanks to this formative assessment activ-ity, the instructor was able to realize this lack of understanding. Instead of giving the

PST the direct answer, the instructor responded to PST's comment with a question in order to make her think about this difference. Again, the formative process did not end here, this point was discussed in the classroom and PSTs were expected to improve their understanding in light of the discussions.

PSTs' analysis reports were full of unclear and short statements and claims without any evidence, as well as statements that showed higher level of thinking. The instructor gave feedback to all of these statements, mostly by questions and requests for further explanations and also by positive statements and check marks. There are many factors in the formative assessment process, some of which are outside the control of the instructor that hinders this process, such as PSTs' nonattendance in some classes, their lack of motivation to participate in the process or their preference to write down short comments. It would be sensible for the instructor to explain this process thoroughly to PSTs before engaging them in it.

Formative assessment is a complex process with many variables that influence its outcome. There are different claims of effectiveness of formative assessment on learning in the literature. Even the concept of "formative assessment" is being discussed by many scholars in terms of its meaning, theory and implementation. A detailed discussion of theoretical aspects of formative assessment is beyond the scope of this chapter. It should suffice to say that despite the ongoing discussions about formative assessment, most scholars agree on its importance and its necessity for implementation (Baird, Andrich, Hopfenbeck, & Stobart, 2017; Hickey, 2015). Accordingly, the formative assessment process discussed here is an example of implementation which was an important part of the teaching and learning process we utilized.

4   Discussion and Implications

This chapter has provided an example of an innovative approach to teach NOS to PSTs, which can also be used in teacher professional development programmes. Helping PSTs to learn and teach about NOS provided them essential knowledge and skills to utilize in their teaching. These skills included argumentation skills such as constructing evidence-based claims, communicating and presenting ideas and embedding NOS teaching within appropriate contexts. This is a first step for PSTs in utilizing NOS teaching skills in other settings. Transferring PSTs skills that they learned in this process in inquiry activities would be a logical next step. Teacher education programmes and professional training programmes based on IBST/L should place enough emphasis on NOS teaching and learning. Research has shown that IBST/L, which only implicitly refer to NOS, is ineffective for learning of NOS (Lederman, 2007). Having grasp of practice as a reasoning resource for inquiry and NOS understanding becomes crucial (Ford, 2009). NOS provides the philosophical foundations for IBST/L and should not be ignored in teacher preparation programmes.

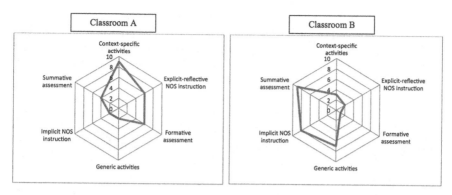

Fig. 2 Radar diagram: implicit vs. explicit-reflective NOS instruction, generic vs. context-specific activities for teaching NOS and summative vs. formative assessment of NOS learning

The pedagogical framework discussed earlier in Fig. 1 and also represented in a radar diagram in Fig. 2 is useful for the organization of the various elements that need to be taken into account during teaching. The radar diagram has three dimensions each varying along a continuum from two aspects: implicit vs. explicit-reflective NOS instruction, generic vs. context-specific activities for teaching NOS and summative vs. formative assessment of NOS learning. We found this radar diagram quite useful while analysing classroom discourse patterns and the nature of teaching. For instance, it allows us to identify what percentage of each dimension is used during a lesson/semester or to compare two different teachers/classrooms. For example, Fig. 2 shows that the teacher in classroom A used more context-specific activities compared to the teacher in classroom B.

The result of our study showed that PSTs' had difficulties in making links between the target aspects of NOS and news in the media reports (Yalaki & Cakmakci, 2011). Therefore, the role of the instructor is crucial for explicitly emphasizing the connections between the target NOS tenets and the news content during the class discussions. The instructor's subject matter knowledge is vital in order to mediate discussions and facilitate PSTs' understanding about the topic (e.g. radiometric dating, human cloning, evolution, nanotechnology, etc.). The instructor needs to know the basic scientific concepts on the topic and if necessary improve his/her related content knowledge before using media reports in classroom.

When opportunities are given to PSTs, they are more likely to raise higher-order questions than questions that simply request declarative information. If instructors wish to utilize this method, they should encourage PSTs to raise questions and use these questions to develop scientific ideas or ideas about NOS. For this purpose, the instructors can use the following questions to encourage PSTs to raise questions or further investigate them: *In order to make an informed decision on this issue, what would you like to know and investigate? After reading this news, what would you like to ask the author of the news and the scientists who have done the research?* Engaging PSTs with this sort of questions has potential to enhance the quality of PSTs' arguments and self-generated questions.

In most cases PSTs effectively used the media report of science in their presentation to peers. It would be interesting to investigate how PSTs transfer this approach into their classrooms and use it when teaching science to their students through inquiry.

During this study, both summative and formative assessments were used to recognize the learners' understanding, and the assessment data were used to inform teaching and to give feedback. Media reports of scientific research were used during formative assessment procedures. PSTs had very positive reactions to the use of formative assessment strategies, and they consistently expressed the usefulness of these strategies for their learning of NOS during several interviews (Yalaki & Cakmakci, 2011).

The strategies for evaluating science-related media news (as presented in Appendix 1) can inform journalists and scientists about the presentation of science-related news. Science journalists may deliver their ideas in a way that enhance public understanding of science and its nature. Strategies that encourage the public to become active participants in, and critical consumers of, science are needed.

Appendix 1: Example 4—Using Media News to Teach about Science Content and NOS

The context: News article and *a video clip*

News article: Finally, Element 117 Is Here! by Lauren Schenkman on 7 April 2010, Science Mag
goo.gl/gzqgd3

Video clip: Video Episode: "The Periodic Table: Mendeleev & Beyond" An interview with Dr. Eric Scerri of UCLA on the history and development of the periodic table.
goo.gl/hp28XB

Aims

The aim of this activity is to bring contemporary science and cutting-edge science into the classroom and explore *the periodic law* and *the structure of atom*. This activity uses a news article as a context to discuss the nature of a scientific law and how science works. The news article is used as a starting point for the development of some scientific ideas in the classroom, including the periodic law and the structure of atom, which may help PSTs to understand the principles behind the periodic table. The following sections consider the use of news articles for supporting the learning of science subject matter knowledge and describe how a news article can be integrated into the science/chemistry curriculum.

Teaching Goals (Nature of Science)
A set of teaching goals are formulated to specify more directly the nature of the pedagogical interventions to be taken by the teacher. Some of these teaching goals are conceptual but some others are epistemological (Leach & Scott, 2002). These specific teaching goals can provide a much more fine-grained analysis of learning points that need to be addressed by the teacher (Leach & Scott, 2002).

This activity can be used to *emphasize* the following NOS tenets:

- Scientific knowledge is based on empirical evidence (NOS-2, see the text).
- Scientific knowledge is partly the product of human inference, imagination and creativity (involves invention of theories and laws) (NOS-5).
- Scientific theories and laws are different kinds of knowledge and serve different functions and that one does not become the other. Generally speaking, theories are inferred *explanations* for observable phenomena, whereas laws are general *descriptions* of the relationships among observable phenomena (Lederman & Abd-El-Khalick, 1998). New evidence supports or disproves a scientific law. As evident in the news article, new research evidence supports the periodic law (NOS-8).

The following ideas may also be emphasized:

- A theory or law should lead to predictions that are precise and detailed enough for it to be possible that it can be shown to be false (AQA, 2010). A theory or law is "scientific" only if it is, among other things, falsifiable, and it is "non-scientific" if it does not make any predictions that could possibly be falsified (Popper, 1959).
- Classification is an important aspect of science. However, Mendeleev contributed to science much more than mere classification; he used his classification scheme (periodic table) to predict the existence of as-yet-undiscovered elements and predicted their properties. The periodic law allowed him to predict undiscovered elements' properties by averaging the characteristics of other elements in the same group.

Teaching Goals (the Science)
This activity may be used to *open up PSTs' own ideas* about atoms, elements, the periodic law and structure of an atom and also to *introduce and support the development of* the ideas related to the periodic table, such as:

- The periodic table itself is a visual representation of the periodic law, which states, "The properties of chemical elements are a periodic function of their atomic numbers".

Certain properties of elements repeat periodically when arranged by atomic number.

Teaching Points and Possible Teaching Sequence
The PSTs will need to have some understanding of concepts such as atoms, elements, molecules and chemical reactions. The teacher may need to encourage the students to use a textbook if they are having difficulty with the definitions of these

concepts. It might be worthwhile to spend some time introducing these concepts. The following teaching sequence is provided as a suggestion for teachers:

1. Distribute the news article to the students and ask students to read the article on their own or alternatively ask a student to read it aloud.
2. During this activity encourage students to comment on the article.
3. Ask students to work in groups to analyse the news article similar to the way described below.
4. After students complete the analysis, a classroom discussion may follow regarding the characteristics of science (NOS tenets) seen in the news.
5. It may not be sufficient just to give students science-related news articles to discuss; the teacher may also need to mediate students' understanding and explicitly address NOS tenets wherever appropriate during the discussion.
6. After the discussion, NOS tenets that clearly stand out in the news should be explicitly emphasized to the class.
7. Introduce the periodic law and give information on the development of this law by Dimitri Mendeleev. A book chapter of Niaz (2008) would be useful to facilitate students' understanding with respect to how scientific progress is laden with controversies, contradictions and alternative interpretations.
8. An interview with Prof. Eric Scerri of University of California, Los Angeles, on the history and the development of the periodic table can be shown to students to reinforce their understanding of the periodic law. Alternatively, students can watch these resources after the class.

    These short video clips are available at: http://elementsunearthed.com/video-episodes

    • In the first part of this video, Prof. Scerri discusses why the periodic table is the central organizing chart of chemistry and how chemists in the early nineteenth century began to organize elements in tables based on atomic weights and properties using the idea of triads. He also discusses Prout's law and the events leading up to the Karlsruhe conference in 1860, a watershed event which led directly to the development of the periodic system. Scientists who developed periodic systems before Dimitri Mendeleev and the reasons for the success of Mendeleev's system are also discussed in this video.
    • In the second part of this video, Prof. Scerri discusses how Dimitri Mendeleev developed his periodic system in 1869 while working on a textbook of inorganic chemistry and how he went on to publish the table and defend it, making bold predictions about missing elements and accommodating over 60 known elements by both atomic weight and chemical properties. His successors continued to revise the table and the discoveries of subatomic physics and quantum mechanics and finally explained the table's structure and the periodic law. He discusses how the discoveries of modern physics such as subatomic particles and quantum mechanics helped us to understand the structure of the periodic table and the properties of the elements. He also discusses whether the periodic table can be derived solely from quantum mechanics and some of the anomalies that remain to be solved, such as the

disputed placement of hydrogen and helium in the periodic table and the mysterious Knight's Move pattern.

9. Conclude that theories and laws are different kinds of knowledge and one does not become the other.

You may also introduce the following web page and suggest students to explore the web page and periodic table. This site is available with several language options, which can be chosen on the right site of the menu: http://www.ptable.com/

Analysing Media Reports of Scientific Research

The news article in this activity is analysed based on the analysis scheme explained in Cakmakci and Yalaki (2011). Please note that it is important to answer the questions in this analysis by referring to the news and/or other sources, even by providing excerpts from the news if necessary. Short answers without detail (such as Yes or No) or without evidence from the news or other sources are discouraged for the sake of argumentation.

Surface Analysis of the News

(a) Do the title, picture and content correlate?
    The picture shows the berkelium produced to make the element 117, which is consistent with the news.
(b) Where is the news published or broadcasted?
    The news is published at the Science Now website.
(c) Is the source reliable?
    Science Now is published by the American Association for the Advancement of Science (AAAS) a well-known and trusted institution in the USA.
(d) What is the circulation rate of the source?
    No specific information is available; however, *Science* magazine and the related websites, including *Science Now*, are well-known and followed publications.
(e) Who wrote the news? Can the original source of the story be identified?
    The news is reported by *Lauren Schenkman*, who has a bachelor's degree in physics and creative writing from the University of Southern California. The research reported in the news appears in *Physical Review Letters*.
(f) Are the results of the scientific research published anywhere else?
    It is mentioned in the news that the results of the study are published in *Physical Review Letters*.
(g) Is there a profit relationship between people and institutions?
    This question cannot be judged from the news.
(h) Who did the research? Who gives their views and how are the scientists involved in the news portrayed? Does the reporter use direct or indirect quotation?
    An international team of scientists from Russia and the USA did the research. The team included scientists from the Joint Institute of Nuclear Research (JINR) (Dubna, Russia), the High Flux Isotope Reactor at Oak Ridge National

Laboratory in Tennessee and Russia's Research Institute of Atomic Reactors in Dimitrovgrad. During this activity, it can be explicitly mentioned that scientists usually work in groups. It was evident in the news that researchers from different countries worked collaboratively to address scientific challenges. The reporter often used direct quotation of scientists' views on the research. A team member Krzysztof Rykaczewski, a nuclear physicist at Oak Ridge, and a nuclear physicist Konrad Gelbke, director of the National Superconducting Cyclotron Laboratory at Michigan State University in East Lansing, can be named.

(i) Are the institutions mentioned in the news reliable?

The institutions mentioned in the news include Joint Institute of Nuclear Research (JINR), Oak Ridge National Laboratory in Tennessee, Russia's Research Institute of Atomic Reactors in Dimitrovgrad and National Superconducting Cyclotron Laboratory at Michigan State University, which seem to be reliable institutions.

(j) Is there a profit relationship between people and institutions?

This question cannot be judged from the news.

By doing this surface analysis, a certain level of trust to the news is established. Now the next phase of the analysis can begin. It should be noted that these questions do not need to be followed strictly, one after another. Rather, these questions are provided as suggestions and students may be allowed to raise and discuss whatever aspects of science and NOS they feel relevant.

Analysis of the News in Relation to NOS Tenets

(a) What are the scientific claims in the news?

It is claimed that the element with atomic number 117 is observed.

(b) What are the evidences that support the claims?

In the news, it is mentioned that scientist bombarded a sample of berkelium with calcium 48 ion for 5 months and recorded the events with detectors and eventually they managed to identify events that provided evidence for the appearance of the element 117. [*Using this example, it can be emphasized that scientific knowledge is empirically based (NOS-2).*]

(c) Are scientific hypotheses, theories or laws mentioned implicitly or explicitly in the news?

The periodic law is implicitly mentioned in the news, which can also be used to explain the different functions of theory and law in science (NOS-8).

(d) Is there information about the scientific methods used in the investigation?

In the news, the procedure that was used in the research is explained, and it is mentioned how the researchers empirically tested their ideas (see paragraphs 1–4 in the news). Using this news, it can also be suggested that scientific knowledge may change in the sense of adding new knowledge to current knowledge (accumulative, evolutionary change). As evident in the news article, new research evidence supports the existing thinking of science (i.e. the periodic law and the notion of the "island of stability") (NOS-1).

It's taken years, but physicists have finally filled in a persistent gap in the periodic table. Eight years after the creation of element 118, the heaviest known atom, researchers have made a few atoms of its slightly lighter neighbor, element 117, by shooting an intense beam of calcium ions into a target of berkelium. Besides sketching in the blank space in the table, the discovery bolsters the notion of an "island of stability", a group of superheavy nuclei still tantalizingly out of reach that theorists predict may be as stable as more familiar elements.

...Experimentally, it's an enormous tour de force", says nuclear physicist Konrad Gelbke, director of the National Superconducting Cyclotron Laboratory at Michigan State University in East Lansing. What's more, "they're developing a picture that's starting to make a lot of sense." Like sailing expeditions of old, the findings are solidifying the existence of an island of stability, made possible through the detailed interactions of neutrons and protons inside the nucleus, he says. "This is decades of very careful and painstaking work that is slowly coming to fruition".

(e) Is there information about the subjects of the investigation?
The subjects of the investigation are various elements.

(f) Is there evidence of a change (or possibility of change) in scientific knowledge?
This news provides a good example of accumulative change in scientific knowledge. It is mentioned in the news that "It's taken years, but physicists have finally filled in a persistent gap in the periodic table. Eight years after the creation of element 118, the heaviest known atom, researchers have made a few atoms of its slightly lighter neighbour, element 117, by shooting an intense beam of calcium ions into a target of berkelium". It is also mentioned that "making element 117 presented a particular challenge", but scientists managed it with a tremendous amount of work.

(g) What are the observations and the corresponding logical deductions?
Scientists observed the events as a result of calcium atoms smashing into berkelium atoms with detectors. From the observed events, they deduced that six atoms of element 117 formed.

(h) Is there evidence of subjectivity among scientists regarding the conclusions of the investigation?
No such evidence is provided in the news.

(i) Does the scientific investigation include creativity and imagination?
The tremendous amount of work and sequence of events explained in the news to synthesize the element 117 presumably include creativity and imagination (NOS-5).

(j) Is there evidence of scientific knowledge being influenced by the social and cultural environment?
This news is an example of an international teamwork among scientists, which is an indication of the social structure of scientific investigation.

(k) Is there evidence of different scientific methods used for the same investigation?
As explained above, many different methods were used to achieve the goal of synthesizing the element 117, which can be given as an example that there is no single scientific method (NOS-3).

(1) What does this research offer to science and society?

The discovery of element 117 provides strong evidence for the existence of the island of stability. Discovery of new elements expands the understanding of the universe, provides important tests of nuclear theories and supports the periodic law. This new discoveries may also trigger other discoveries.

Analysis of the News in Relation to Science Concepts and Curriculum

This news article can be used as a context to teach scientific concepts related to periodic table and the periodic law. However, in more advanced physics or chemistry courses, ideas such as the island of stability could also be introduced.

What Science Says?

In 1871, the Russian chemist Dimitri Ivanovich Mendeleev (1834–1907) proposed the periodic law. Mendeleev arranged the elements in order of increasing relative atomic mass, and his periodic law stated that "the properties of the elements are a periodic function of their relative atomic masses". While constructing this table, Mendeleev found that there were not enough elements (at that time about 60 elements were known) to fill all the available space in each horizontal row or period. He assumed that eventually these elements would be discovered in the future. Therefore, he left blank spaces for undiscovered elements and predicted their properties by averaging the characteristics of other elements in the same group. While Mendeleev's periodic law allowed him to predict the behaviour of elements, this law does not explain why it happened. There seems to be considerable controversy among philosophers of science with respect to the nature of Mendeleev's periodic law (Niaz, 2008). For instance, Niaz, Rodrguez, and Brito (2004) argue, "despite Mendeleev's own ambivalence, periodicity of properties of chemical elements in the periodic table can be attributed to the atomic theory.... Mendeleev's contribution can be considered as an 'interpretative' theory which became 'explanatory' after the periodic table was based on atomic numbers". Scerri and Worrall (2001) claim, "the Periodic Table is patently not itself a theory and therefore does not in itself have any logical consequences. Mendeleev saw his Table (indeed, significantly, Tables—he produced a total of sixty-five different ones through the course of his career) as embodying, or as underpinned by, something he called the 'periodic law'". Molecular orbital theory and theories in quantum mechanics offer possible explanations for such behaviours (Scerri, 2006). Mendeleev's periodic law is modified by these theories, and the modern form of the periodic law states that "the properties of chemical elements are a periodic function of their atomic numbers". Certain properties of elements repeat periodically when arranged by atomic number. For example, progressing from left to right across the modern periodic table (wide form), certain properties of the elements approximate those of precursors at regular intervals of 2, 8, 18 and 32. For example, the 2nd element (helium) is similar in its chemical behaviour to the 10th (neon), as well as to the 18th (argon), the 36th (krypton), the 54th (xenon) and the 86th (radon) (see the extreme right column in the modern periodic table-wide form). The chemical family called the halogens, composed of elements fluorine (atomic number = 9), chlorine (17), bromine (35), iodine (53) and astatine (85), and finally ununseptium, element 177 (177), is an extremely reactive family.

Concluding Remarks

This news article can be used as a context to teach different scientific concepts and aspects of NOS; however, we used the news as a starting point to introduce and discuss ideas about periodic table and the periodic law. This activity can also be used in a chemistry lesson while teaching about the periodic table. A book chapter on the periodic table written by Niaz (2008) can give the reader some thought-provoking ideas.

References

Abd-El-Khalick, F., & Lederman, N. G. (2000). The influence of history of science courses on students' views on nature of science. *Journal of Research in Science Teaching, 37*(10), 1057–1059.

AQA. (2010). *GCE-AS and a level specification, Science in society.* Available at: http://store.aqa.org.uk/qual/gce/pdf/AQA-2400-W-SP.PDF

Baird, J., Andrich, D., Hopfenbeck, T. N., & Stobart, G. (2017). Assessment and learning: Fields apart? *Assessment in Education: Principles, Policy & Practice, 24*(3), 317–350.

Black, P., & Wiliam, D. (1998). Assessment and classroom learning. *Assessment in Education: Principles, Policy & Practice, 5*(1), 7–73.

Cakmakci, G. (2012). Promoting pre-service teachers' ideas about nature of science through educational research apprenticeship. *Australian Journal of Teacher Education, 37*(2), 114–135.

Cakmakci, G., & Yalaki, Y. (2011, July 1–5). *Using media reports of scientific research as a medium for teaching science and nature of science.* In The International History, Philosophy, and Science Teaching (IHPST) conference, Thessaloniki, Greece.

Cakmakci, G., & Yalaki, Y. (2012). *Promoting student teachers' ideas about nature of science through popular media.* Trondheim, Norway: S-TEAM/NTNU.

CERI. (2005). *Formative assessment: Improving learning in secondary classrooms.* Paris: OECD.

Clough, M. P. (2006). Learners' responses to the demands of conceptual change: Considerations for effective nature of science instruction. *Science Education, 15*, 463–494.

CNN Turk. (2010). *Cep telefonları kanser yapıyor mu?* Retrieved May 18, 2010, from https://goo.gl/2knVsq. Full English translation of the news can be found at: https://goo.gl/BWWg43

Conant, J. B. (1951). *Science and common sense.* New Haven, CT: Yale University Press.

Cowie, B., & Bell, B. (1999). A model of formative assessment in science education. *Assessment in Education, 6*(1), 101–116.

Cumhuriyet. (2010). *İlk modern insan 400 bin yil once mi cikti?* Retrieved March 01, 2011, from https://goo.gl/BsFZtv Full English translation of the news can be found at: https://goo.gl/bBMXvT

Driver, R., Leach, J., Millar, R., & Scott, P. (1996). *Young people's images of science.* Buckingham, UK: Open University Press.

Duschl, R. A. (2000). Making the nature of science explicit. In R. Millar, J. Leach, & J. Osborne (Eds.), *Improving science education: The contribution of research* (pp. 187–206). Philadelphia: Open University Press.

Elliott, P. (2006). Reviewing newspaper articles as a technique for enhancing the scientific literacy of student teachers. *International Journal of Science Education, 28*(11), 1245–1265.

Ford, D. J. (2009). Promises and challenges for the use of adapted primary literature in science curricula: Commentary. *Research in Science Education, 39*(3), 385–390.

Harrison, C. (2015). Assessment for learning in science classrooms. *Journal of Research in STEM Education, 1*(2), 78–86.

Hickey, D. T. (2015). A situative response to the conundrum of formative assessment. *Assessment in Education: Principles, Policy & Practice, 22*(2), 202–223.

Jarman, R., & McClune, B. (2007). *Developing scientific literacy: Using news media in the classroom.* Maidenhead, UK: Open University Press.

Khishfe, R., & Abd-El-Khalick, F. (2002). Influence of explicit and reflective versus implicit inquiry-oriented instruction on sixth graders' views of nature of science. *Journal of Research in Science Teaching, 39*, 551–578.

Leach, J., Hind, A., & Ryder, J. (2003). Designing and evaluating short teaching strategy about the epistemology of science in high school classroom. *Science Education, 87*(6), 831–848.

Leach, J., & Scott, P. (2002). Designing and evaluating science teaching sequence: An approach drawing upon the concept of learning demand and a social constructivist perspective on learning. *Studies in Science Education, 38*, 115–142.

Lederman, N. (2006). Syntax of nature of science within inquiry and science instruction. In L. Flick & N. Lederman (Eds.), *Scientific inquiry and nature of science: Implications for teaching, learning, and teacher education* (pp. 301–317). Boston: Kluwer.

Lederman, N. G. (1999). Teachers' understanding of the nature of science and classroom practice: Factors that facilitate or impede the relationship. *Journal of Research in Science Teaching, 36*(8), 916–929.

Lederman, N. G. (2007). Nature of science: Past, present and future. In S. A. Abell & N. G. Lederman (Eds.), *Handbook of research on science education* (pp. 831–879). London: Lawrence Erlbaum Associates.

Lederman, N. G., & Abd-El-Khalick, F. (1998). Avoiding de-natured science: Activities that promote understandings of the nature of science. In W. McComas (Ed.), *The nature of science in science education: Rationales and strategies* (pp. 83–126). Dordrecht, The Netherlands: Kluwer.

Lederman, N. G., Abd-El-Khalick, F., Bell, R. L., & Schwartz, R. S. (2002). Views of the nature of science questionnaire: Toward valid and meaningful assessment of learner's conceptions of the nature of science. *Journal of Research in Science Teaching, 39*(6), 497–521.

Linn, M. C., Davis, E. A., & Bell, P. (2004). Inquiry and technology. In M. C. Linn, E. A. Davis, & P. Bell (Eds.), *Internet environments for science education*. Mahwah, NJ: Lawrence Erlbaum Associates.

McComas, W. F. (2017). Understanding how science works: The nature of science as the foundation for science teaching and learning. *The School Science Review, 98*(365), 71–76.

MEB (Turkish Ministry of National Education). (2018). *Science lessons curriculum* (grades 3-8). Retrieved from http://mufredat.meb.gov.tr/

Millar, R. (2006). Twenty first century science: Insights from the design and implementation of a scientific literacy approach in school science. *International Journal of Science Education, 28*(13), 1499–1521.

Minner, D. D., Levy, A. J., & Century, J. R. (2010). Inquiry-based science instruction – What is it and does it matter? Results from a research synthesis years 1984 to 2002. *Journal of Research in Science Teaching, 47*, 474–496.

National Research Council. (2000). *Inquiry and the National Science Education Standards.* Washington, DC: The National Academies Press.

NGSS Lead States. (2013). *Next generation science standards: For states, by states ( ).* Retrieved from http://www.nextgenscience.org/

Niaz, M. (2008). *Physical sciences textbooks: History and philosophy of science.* New York: Nova Science Publishers.

Niaz, M., Rodrguez, A., & Brito, A. (2004). An appraisal of Mendeleev's contribution to the development of the periodic table. *Studies in History and Philosophy of Science, 35*, 271–282.

Nicol, D. J., & Macfarlane-Dick, D. (2006). Formative assessment and self-regulated learning: A model and seven principles of good feedback practice. *Studies in Higher Education, 31*(2), 199–218.

Norris, S. P., Phillips, L. M., & Korpan, C. A. (2003). University students' interpretation of media reports of science and its relationship to background knowledge, interest and reading difficulty. *Public Understanding of Science, 12*, 123–145.

Osborne, J. F., Erduran, S., & Simon, S. (2004). Enhancing the quality of argument in school science. *Journal of Research in Science Teaching, 41*(10), 994–1020.

Pedaste, M., Maeots, M., Siiman, L. A., de Jong, T., van Riesen, S. A. N., Kamp, E. T., et al. (2015). Phases of inquiry-based learning: Definitions and the inquiry cycle. *Educational Research Review, 14*, 47–61.

Popper, K. (1959). *The logic of scientific discovery*. London: Routledge.

Ratcliffe, M., & Grace, M. (2003). *Science education for citizenship: Teaching socio-scientific issues*. London: McGraw-Hill Education.

Sadler, T. D. (2006). Promoting discourse and argumentation in science teacher education. *Journal of Science Teacher Education, 17*(4), 323–346.

Scerri, E., & Worrall, J. (2001). Prediction and the periodic table. *Studies in History and Philosophy of Science, 32*, 407–452.

Scerri, E. R. (2006). *The periodic table: Its story and its significance*. New York: Oxford University Press.

Seckin Kapucu, M., Cakmakci, G., & Aydogdu, C. (2015). The influence of documentary films on 8th grade students' views about nature of science. *Educational Sciences: Theory & Practice, 15*(3), 797–808.

Storksdieck, M. (2016). Critical information literacy as core skill for lifelong STEM learning in the 21st century: Reflections on the desirability and feasibility for widespread science media education. *Cultural Studies of Science Education, 11*(1), 167–182.

Yalaki, Y., & Cakmakci, G. (2011, July 1–5). *Formative assessment to enhance students' learning of nature of science*. In The International History, Philosophy, and Science Teaching (IHPST) Conference, Thessaloniki, Greece.

The Development of Collaborative Problem-Solving Abilities of Pre-service Science Teachers by Stepwise Problem-Solving Strategies

Palmira Pečiuliauskienė and Dalius Dapkus

1 Introduction

The complex, multidimensional world we live in requires people to make connections among various elements of knowledge in order to adapt to their environment, develop and act effectively in it. The ability to create new intellectual products while working in collaboration is quite important in the creative society (Florida & Tinagli, 2004). Labour market experts, managers of human resources and vocational education and training experts consider these abilities to be essential. A 2010 Eurobarometer survey showed that significant numbers of employers questioned claimed that the ability to work well in a team (98%) and to adapt to new situations (97%) and communication skills (96%) were important when being recruited for their companies (Flash Eurobarometer reports, 2010). It means that leaders in the business and public sectors highly valued employees' team working abilities, while these employers also thought that analytical skills, problem solving and adaptation to new situations were important.

Problem-based learning (PBL) is a good way for preparing people for problem solving in everyday practice because problem solving is generally regarded as the most important cognitive activity in everyday and professional practice (Pierrakos, Anderson, & Barrella, 2016). PBL learners maximize learning with investigation, explanation and resolution by starting from real and meaningful problems (Oguz-Unver & Arabacioglu, 2014). In PBL learners solve problems based on their prior knowledge and experience.

Therefore, PBL is often missed in educational practice. Gok (2014) states that in the classroom, teachers often teach concepts, principles and formulas regarding the

P. Pečiuliauskienė (✉) · D. Dapkus
Department of Biology and Chemistry, Lithuanian University of Educational Sciences,
Vilnius, Lithuania
e-mail: palmira.peciuliauskiene@leu.lt; dalius.dapkus@leu.lt

© Springer International Publishing AG, part of Springer Nature 2018   163
O. E. Tsivitanidou et al. (eds.), *Professional Development for Inquiry-Based
Science Teaching and Learning*, Contributions from Science Education
Research 5, https://doi.org/10.1007/978-3-319-91406-0_9

course subjects and then students conventionally solve several sample problems and do not reflect their success. Students cannot develop any systematic problem-solving strategies in this way. "All learning involves active thinking, and instructors should create more room for their learners to construct their own knowledge" (Oguz-Unver & Arabacioglu, 2014, p. 120).

Scholars (Gok, 2014; Pólya, 1945; Reif, Larkin, & Brackett, 1976) propose a systematic problem-solving strategies way. The researchers distinguish the following three stepwise SPS strategies: (1) identification (this stage means identification of fundamental problems), (2) solution (this stage in essence involves implementation) and (3) checking (this stage mainly comprises monitoring, setting and controlling). Gok (2014) examined the effects of SPS on students' achievement, skill and confidence. He states that "Problem solving strategy steps including conceptual learning, solution, and crosscheck are proved to be statistically effective in problem solving" (Gok, 2014, p. 618).

Others scholars (Moskovitz & Kellog, 2011; Pierce, 2008; Savery, 2006;Welch, Klopfer, Aikenhead, & Robinson, 1981) describe the peculiarities of learning activity at different SPS strategies. At the initial stages of PBL, during the identification of the problem (identification strategy), observational skills are identified as having a high priority. According to Savery (2006), the problem-solving element of PBL requires learners to look at the multiple perspectives of problems. Learners should be able to access, study and integrate information from different disciplines. Multiple and cross-curricular perspectives lead learners to a more thorough understanding of issues and the development of a more robust solution through PBL (Savery, 2006). Learners should understand and explain what they are learning (solution and checking strategies) (Moskovitz & Kellog, 2011; Pierce, 2008; Welch et al., 1981).

PBL is a constructivist pedagogical approach to learning in which learners work together to find solutions to a complex problem (Ferreira & Trudel, 2012). Problems used in PBL must be ill-structured (or complex) and allow free inquiry (Savery, 2006). The PBL process can construct an extensive, flexible and multidisciplinary knowledge base. This is related to the cross-curricular content of PBL because learners can incorporate prior knowledge of complex problems from different subjects (Oguz-Unver & Arabacioglu, 2014).

Cross-curricular problems create conditions for learners to transfer their knowledge from one subject to another in every cross-curricular learning situation (horizontal shift) (Funke, 1991; Hunt, 1994; Savin-Baden, 2016). The shift of knowledge from one subject into the contents of the teaching situation of another subject provides new character, creates problem situations, encourages learners to acquire new information or envisages new aspects of the knowledge acquired (Dörner & Funke, 2017; Edelson, 2001; Savin-Baden, 2005; Zoller, 2011). However, cross-curricular educators have very little information on how to implement problem-based teaching in classrooms where multiple disciplines are represented (Keebaugh, Darrow, Tan, & Jamerson, 2009). The S-TEAM project has contributed to this movement by adopting PBL. Our participation in the S-TEAM project encouraged us to look for the possibilities for the improvement of PBL and simultaneously find solutions to cross-curricular problems and collaborative learning based on problem-solving strategies.

In cross-curricular problems solving, students work together in mixed groups seeking to achieve common goals; and for this, they have to discuss with each other, as well as help each other (Doymuş, Şimşek, & Bayrakçeken, 2004; Johnson & Johnson, 1999). Collaborative problem solving is "the capacity of an individual to effectively engage in a process whereby two or more agents attempt to solve a problem by sharing the understanding and effort required to come to a solution and pooling their knowledge, skills and effort to reach that solution" (OECD, 2013).

According to Hansson, Foldevi, and Mattsson (2010), collaboration and teamwork probably have to be experienced all through the curriculum and need to be integrated into most of the teaching if they are to have a more evident effect. Scholars (Kumar & Natarajan, 2007; Schmidt, Van der Molen, Te Winkel, & Wijnen, 2009) reveal that when students collaborate in the solution of a problem, one of the most important outcomes of PBL is the development of interpersonal abilities (Kumar & Natarajan, 2007; Schmidt et al., 2009).

Collaborative problem-solving abilities of school students depend on teachers' abilities to promote collaborative problem-solving abilities of students. So CPS abilities have to be developed during pre-service teachers' education. None of the studies reviewed have examined the impact of PBL by stepwise problem-solving strategies on the collaborative problem-solving abilities of pre-service science teachers. Therefore, complex attitudes towards the application of the models, as well as educational insights into their coherence in educational practice, especially in training pre-service science teachers, are still lacking. It is important to reveal the impact of SPS towards pre-service science teachers' collaborative problem-solving abilities.

The aim of the research is to reveal the role of collaborative stepwise problem-solving strategies in the development of cross-curricular problem-solving abilities of pre-service science teachers.

The research questions are as follows:

1. What are the collaborative problem-solving abilities of pre-service science teachers at problem identification and problem implementation stages of SPS?
2. How are collaborative problem-solving abilities of pre-service science teachers related with the attitude to use cross-curricular relationships and collaborative learning in the future educational practice at school?
3. What are the attitudes of pre-service science teachers about the role of the cross-curricular content projects in the concretization of a problem, prediction of a problem-solving scenario, management of information and monitoring of problem solving?

2 Theoretical Background

The question concerning the development of collaborative problem-solving abilities by SPS strategies cannot ignore inquiry-based learning (IBL) tradition. During IBL, learners acquire knowledge from direct observations by using deductive questions. Inquiry-based learning (IBL) is the framework for PBL and problem solving. "IBL focuses on knowledge construction and, taking account knowledge transference, IBL gives way to PBL" (Oguz-Unver & Arabacioglu, 2014, p. 127). IBL is the intentional process of diagnosing problems, critiquing experiments, distinguishing alternatives, planning investigations, researching conjectures, searching for information, constructing models, debating with peers and forming coherent arguments (Linn, Davis, & Bell, 2004). IBL can be realized using different strategies, methods and contents associated with problem-based learning situations (Capon & Kuhn, 2004; Dochy, Segers, Bossche, & Gijbels, 2003; Gallagher, Stepien, & Rosenthal, 1992; Harlen & Allende, 2006; Mayer, 2004; Moskovitz & Kellog, 2011).

IBL and PBL have commonalities and differences (Savery, 2015). On the one hand, IBL and PBL are very similar because both are grounded in the philosophy of John Dewey and use constructivist student-centred approach. On the other hand, PBL and IBL are different because at PBL students solve problems based on their prior knowledge and experience, as well as key elements of IBL – exploration, invention and application (Oguz-Unver & Arabacioglu, 2014; Savery, 2006). Both IBL and PBL involve problem-solving activity that requires effort to achieve a certain goal to eliminate the encountered difficulties. The learners who have gained problem-solving abilities can overcome simple or complex problems faced by the society in our rapidly changing environment.

PBL and IBL allow learners to project their learning process. On the other side, it is necessary to have very good abilities of planning the learning process. In other words, the creation of the model based on constructivist collaborative learning and cross-curricular content has to be considered so that the advantage of guidance begins to recede only when learners have sufficiently high prior knowledge to provide "internal" guidance (Kirschner, Sweller, & Clark, 2006).

IBL is used in science education and is named inquiry-based science learning (IBSL). IBSL has been proposed as a framework for conceptualizing the priorities and values of authentic science learning and encourages hands-on approach "where students practice the scientific method on authentic problem questions" (Savery, 2015, p.11).

IBSL is characterized by a variety of levels. We refer to the theory of Banchi and Bell (2008) when deciding how to teach pre-service teachers. The lowest level of IBSE (confirmative inquiry) corresponds to the activities where learners know the possible outcomes of problem solving and where a detailed description of activities and problems is provided (Table 1).

The second level of IBSE (structured inquiry) is reached in the projects when learners are provided with a problem and the method for its solution. The third level (coordinated inquiry) is characterized by the fact that learners know the problem but

Table 1 Levels of inquiry-based learning

Level of inquiry	Question/problem	Procedure	Solution
Confirmative inquiry	+	+	+
Structured inquiry	+	+	−
Coordinated inquiry	+	−	−
Open inquiry	−	−	−

According to Banchi and Bell (2008)

have to find out how to solve it by themselves. The highest level (open inquiry) is reached when learners identify a problem, methods for its solution and explanations for the cross-curricular phenomena themselves. Therefore, PBL based on cross-curricular content corresponds to the open inquiry, as cross-curricular problems activate the knowledge of different subjects and create conditions for learners to transfer their knowledge from one subject to another in every cross-curricular learning situation.

According to Metallidou (2009), problem solving as a goal-directed behaviour requires an appropriate mental representation of the problem and the subsequent application of certain methods or strategies in order to move from an initial – current state to a desired – goal state. Problem solving is a continuous process consisting of the following three SPS strategies: identification, solution and checking (Gok, 2014; Pólya, 1945; Reif et al., 1976). Bransford and Stein (1984) propose two types of problem-based learning models: progressive and cyclical. According to the application of the progressive model, greater attention is paid to the solution of a problem and its application in practice (Bransford & Stein, 1984). In the cyclical model, greater attention is paid to the definition of a problem, its analysis and search for information (Boud & Feletti, 1997). The latter model deals with the recognition of a problem and its continuous revision.

Various studies (Gok, 2014; Hayes, 1989; Jonassen, 2011; Pretz, Naples, & Sternberg, 2003) focus on how students learn in the different phases of the PBL cycle. According to Gok (2014), at the first cycle, learners should comprehend concept(s)/principle(s), determine the known and unknown variables, visualize problems in the light of their own knowledge and restate the problem in their own words in the first step of problem solving. The second cycle involves implementation (qualitative and quantitative problem solving). In the last problem-solving cycle (checking), learners should check the solution and explore alternative ways of solving a problem. Jonassen (2011) indicates a similar problem-solving structure. Hayes (1989) distinguishes seven phases of problem solution: identification, recognition, definition, presentation of a problem, creation of problem-solving strategy, structuration of knowledge necessary for problem solution, attribution of psychical and physical resources, monitoring of problem solution process and evaluation of problem solution.

According to Sternberg and colleagues (Pretz, et al., 2003, p. 4–5), problem-solving process can be described as a cycle of seven steps or events: (1) a problem is recognized or identified in the environment; (2) the problem is defined and represented mentally; (3) within the mental representation generated, a solution strategy is developed to solve the problem; (4) relevant knowledge about the problem is organized; (5) the physical and mental resources needed to solve the problem are distributed; (6) progress towards the goal of solving the problem is monitored; and (7) the solution is evaluated for meeting the goal of solving the problem.

Foldevi (1995) analyses PBL in medical schools and distinguishes nine steps of problem solving: creation of a problem scenario, creation of a group plan, formulation of a hypothesis of problem solution, "brainstorming", definition of the problem, formulation of learning tasks, deepening one's knowledge, discussions and careful research of knowledge and application of knowledge in practice.

The review of various studies (Foldevi, 1995; Gok, 2014; Hayes, 1989; Jonassen, 2011; Pretz, et al., 2003) reveals more commonalities rather than differences how students learn in the different phases of the PBL cycle. Some authors (Pretz, et al., 2003) specify seven; others (Foldevi, 1995) point out nine steps. Although the number of steps in the PBL is different, learning activities are similar and are based on learners' prior knowledge and experience.

3   Research Methodology

3.1   Participants of the Research

Pre-service teachers (third year students) studying at different science study programmes (Biology, Chemistry and Physics) at Lithuanian University of Educational Sciences were involved in the research. The authors of this paper organized joint seminars, and pre-service teachers worked in mixed groups of six people. Each group worked exclusively on cross-curricular problems within short-term projects.

The pre-service teachers had to create models of cross-curricular relationships, predict possible links and suggest methods for their analysis during a lesson. For example, they had to present a theme "Atmosphere: its physical and chemical aspects". From the point of view of Chemistry, it was important to disclose the structure of atmosphere, as well as the formation of oxygen and oxidation of metals. From the point of view of Biology and Physics, pre-service teachers had to disclose the importance of oxygen to life (photosynthesis, respiration, etc.), as well as the formation of atmosphere and its layers, etc.

The realization of a cross-curricular project requires good understanding of content of different science subjects. Therefore, it is quite difficult to realize it in practice, e.g. learners of Physics know the content of Physics, but their knowledge of Chemistry or Biology is much weaker (Keebaugh et al., 2009). It is much easier to disclose possible cross-curricular relationships when pre-service teachers of Biology, Physics and Chemistry form heterogeneous groups and work together. In order to prepare favourable conditions for the collaborative learning, the time schedules of seminars for science education learners were synchronized. The pre-service teachers implemented one cross-curricular project in two seminars (four academic hours). They worked independently (three academic hours) between the seminars. Self-evaluation of pre-service teachers was performed individually at the end of each SPS stage of the problem-solving process.

The group of pre-service teachers for the quantitative research was composed of 120 respondents (50% biologists, 25% chemists, 25% physicists). The qualitative research was based on a semi-structured interview of 18 purposefully chosen pre-service teachers from Physics (three pre-service teachers), Biology (three pre-service teachers) and Chemistry groups (three pre-service teachers). The sample of the qualitative research was typical case sampling (cases that are not in any way atypical, extreme, deviant or unusual) (Patton, 2002). Identifying typical cases helped us to identify and understand the key aspects of a CPS phenomenon as they are manifested under ordinary circumstances in science education.

3.2   Instruments of the Research

Quantitative questionnaires and qualitative (semi-structured interview) methods were used in this case study.

In the quantitative research, we used a questionnaire of two parts: (1) collaborative problem-solving processes and (2) factors affecting CPS (intended educational practice, socio-demographic data).

Collaborative problem solving includes understanding and representing the problem content, applying problem-solving strategies and applying self-regulation and metacognitive processes to monitor progress towards the goal (Funke, 2010). The first part of the questionnaire was made on the basis of Social Problem-Solving Inventory (SPSI). We adapted Social Problem-Solving Inventory (SPSI) for the measurement of CPS. SPSI is divided into four subscales: the problem definition and formulation subscale, the generation of alternative solutions subscale, the decision-making subscale and the solution implementation and verification subscale. The pre-service teachers completed questionnaires after each cross-curricular project. We collected 480 questionnaires because we had 120 respondents and 4 cross-curricular projects ($120 \times 4 = 480$).

Every question in the adapted questionnaire had an interval scale from 1 to 100 scores. Questionnaires that are going to yield numerical data can be analysed using statistic and computer programmes (e.g. SPSS). The reliability of adapted questionnaires for Lithuanian pre-service teachers was verified by assessing CPS abilities. The results of Cronbach alpha values were 0,759.

The qualitative research was based on semi-structured interviews. Questions about CPS abilities on cross-curricular content were prepared in advance and divided into five groups: collaborative learning, prediction of a problem-solving scenario, concretization of a problem, management of information and monitoring. The reliability of the interview tended to be obtained by inter-rater reliability: analysing whether other teachers of science education (biologists, chemists and physicists) with the same theoretical framework and observing CPS abilities on cross-curricular content phenomena interpreted them in the same way. We used interview, with the same format and sequence of words and questions for each focus group.

4   The Context of the Research: Description of the Learning Method

The PBL model corresponding to the open inquiry-based concept was created during the S-TEAM project. Collaborative learning based on the cross-curricular relationship model was chosen because it matched the highest level of inquiry (Banchi & Bell, 2008). The role of teachers of science education (tutors) was minimal while working according to the PBL model. In open inquiry, a tutor provides the learners with research questions only, and learners design the procedure (method) to test their questions and the resulting explanations themselves. We created an educational model which was based on four dimensions: learning content, learning concept, learning method and learning result (Fig. 1).

As it was mentioned earlier, PBL is based on (SPS) strategies: (1) identification (this stage means identification of fundamental problems), (2) solution (this stage in essence involves implementation) and (3) checking (this stage mainly comprises monitoring, setting and controlling). Each stage has its own distinct and important contribution towards the discovery of an effective or adaptive solution.

The goal of problem identification is to obtain relevant and factual information about the problem, clarify the nature of the problem and delineate a set of realistic problem-solving goals. At the first step of identification, learners discuss the problem statement and explore the issues. Learners feel that they do not know enough to solve the cross-curricular problems (pre-service teachers of Physics lack the knowledge of Biology and Chemistry; pre-service teachers of Biology lack the knowledge of

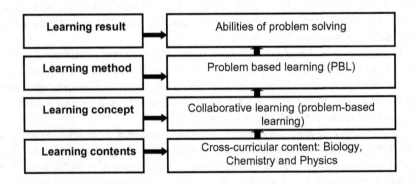

Fig. 1 Learning components of PBL regarding collaborative learning on the basis of cross-curricular content

Physics and Chemistry, etc.). At the second step of identification of a problem, deepening the knowledge, learners analyse what they know to solve the cross-curricular problem, as well as what strengths and capabilities each team member has.

At the third step, learners define a problem (develop and write out the problem statement in their own words). At the fourth step, they formulate a hypothesis of problem solving. Formulation of the hypothesis provides a causal explanation or proposes some association between the elements of cross-curricular content. Carrying out cross-curricular content projects learners perform their ability to formulate hypotheses about realistic cross-curricular situations.

At the fifth step (brain storming), learners look for cross-curricular problem solving. The generation of alternative solutions is to identify, discover or create as many solution alternatives as possible in such a way as to maximize the likelihood that the best possible solution to the cross-curricular problem or solutions is among them. One should look at cross-curricular problems in different ways (from the viewpoints of a physicist, biologist or chemist) and find a new perspective that one has not thought of before.

At the sixth step (concretization) of problem solving, the problem statement can be revised and edited as new information is discovered or "old" information is discarded. In the concretization of the problem (decision-making), the objective is to judge and compare the different alternatives and choose the best overall solution for the implementation within the actual cross-curricular problem-based situation.

At the seventh and the eighth steps, learners form a strategy for cross-curricular problem solution. They create a cross-curricular problem-solving scenario. At the eighth step, learners create a group plan (discuss possible resources, books and websites, and assign and schedule research tasks, especially deadlines).

The second stage of SPS involves the implementation of cross-curricular problem solving: management of information, explanation of problem solution as well as practical application of experience. At the last stage of SPS, learners take pride in what they have done well and learn from what they have not done well.

The learning content in our project was cross-curricular: Biology, Chemistry and Physics. The search for cross-curricular relationships was an active and open process in which individual learners could interpret connections between facts and could find specific relationships among the contents of Biology, Chemistry and Physics. This diversity of links was more easily disclosed when the same project was implemented in a group of pre-service teachers studying at different science programmes. The diversity of cross-curricular relationships, therefore, enhanced the aspects of collaborative learning, including the ability to achieve a single solution and the ability to choose similar models of cross-curricular relationships and problem-solving scenarios.

The list of problem-solving abilities (Table 2) was made on the basis of the overview of theoretical background for the process of problem-based learning (PBL) and the studies examining its effectiveness (Foldevi, 1995; Gok, 2014; Hayes, 1989; Jonassen, 2011; Pólya, 1945; Pretz et al., 2003). Various problem-solving abilities were unfolded in each learning phase (Table 2). Different abilities were developed during different steps of educational practice.

Table 2 Problem-solving
abilities and SPS strategies

Problem-solving abilities	SPS strategies
Discussions and careful research of knowledge	Identification
Deepening of one's knowledge	
Definition of the problem	
Formulation of a hypothesis of problem solution	
"Brainstorming"	
Concretization of the problem	
Creation of a problem-solving scenario	
Creation of a group plan	
Management of information	Implementation
Explanation of problem solution	
Application of knowledge in practice	

The role of teachers of science education (tutors) was limited to the presentation of topics of cross-curricular projects during the realization of open learning project activity. The tutors of Biology, Chemistry and Physics provided some topics while creating modules of didactics of cross-curricular teaching. The modules were based on comparative tables of concepts, theoretical analysis of cross-curricular relationships and practical tasks for their realization. The preparation of the modules encouraged cooperation of teacher tutors of science education. The topics for the PBL projects for pre-service teachers were selected according to the national general programmes of elementary and basic education (General Programmes of Elementary and Basic Education, 2008) and national science textbooks. Four cross-curricular topics were presented during each seminar for pre-service teachers (the example of topic list for one seminar: E-textiles in our life: protects or hurts?; Silence in a big city: how to increase it?; Solar battery in our life: is it worth building solar cell paths?; Smart phone: how to create an eco-phone?). Besides, the learners had a possibility to choose topics of a project by themselves.

5   Results

5.1   The Results of Quantitative Research

The results of the research were analysed by stepwise problem-solving strategies based on the three stages of SPS: identification, implementation and checking (Table 3). The results of identification and implementation were based on quantitative research (Tables 4, 5, and 6), whereas the results of monitoring and evaluation (checking) were based on quantitative and qualitative data analysis (Tables 7 and 8).

Eight steps of problem-solving activities (discussions and careful research of knowledge, deepening of one's knowledge, definition of the problem, formulation

Table 3 Problem-solving activities, stepwise problem-solving (SPS) strategies and research

Learning activities	SPS strategies	Research
Discussions and careful research of knowledge	Identification	Quantitative
Deepening of one's knowledge		
Definition of the problem		
Formulation of a hypothesis of problem solution		
"Brainstorming"		
Concretization of the problem		
Creation of a problem-solving scenario		
Creation of a group plan		
Management of information	Implementation	Quantitative
Explanation of problem solution		
Application of knowledge in practice		
Monitoring and evaluate	Checking	Qualitative

Table 4 Problem-solving abilities of pre-service teachers: descriptive statistics data

Problem-solving abilities	SPS strategies	Mean	Standard deviation
1. Discussions and careful research of knowledge	Identification of problem	56.89	11.78
2. Deepening of one's knowledge		55.14	11.58
3. Definition of the problem		49.12	19.24
4. Formulation of a hypothesis		64.22	13.11
5. Brainstorming		59.31	12.47
6. Concretization of the problem		55.11	12.17
7. Creation of a problem-solving scenario		61.12	14.10
8. Planning of problem solution		65.44	13.23
9. Management of information	Implementation of problem	69.14	15.45
10. Explanation of problem solution		73.25	18.87
11. Practical application of experience		72.24	17.65

of hypothesis of problem solution, "brainstorming", concretization of the problem and creation of a group plan) are related to the first SPS strategy named "Identification of a problem". Three steps (management of information, explanation of problem solution, practical application of experience) of problem-solving activities are related to the second SPS strategy named "Implementation of problem solution" (Table 3). The monitoring and evaluation at the end of each strategy are also important attributes of the PBL process.

Self-evaluation of pre-service teachers of CPS abilities was performed at the end of each stage. The mean of self-evaluation of problem-solving abilities was calculated after four cross-curricular educational projects. The results (Table 4) of our research showed that the means of problem-solving abilities that were based on self-evaluation of pre-service teachers varied from 49,12 to 73,25 scores. The means

Table 5 Results of ANOVA Bonferoni data block test. Pairwise comparisons between problem-solving abilities of pre-service teachers

(I) factor1	(J) factor 1	Mean difference (I–J)	Std. error	Significance[a]	95% confidence interval Lower bound	Upper bound
Explanation of problem	1. Discussion	15.784	2.324	0.000	1.780	18.745
	2. Deepening	15.115	2.105	0.000	−1.401	18.531
	3. Planning	10.727*	2.628	0.001	3.048	16.405
	4. Formulation (hypothesis)	−11.115	2.224	0.031	−7.231	13.405
	5. Brainstorming	12.247	2.411	0.023	2.323	14.225
	6. Concretization	17.722*	2.623	0.000	1.297	14.128
	7. Definition	19.627*	2.424	0.000	−16.308	−5.057
	8. Creation of scenario	12.312	2.171	0.045	8.451	10.451
	9. Management	−5.712*	2.512	0.514	−13.128	−2.397
	10. Application	3.215	2.213	0.638	−12.401	7.432

Based on estimated marginal means
*The mean difference is significant at the 0.05 level
[a]Adjustment for multiple comparisons: Bonferroni

Table 6 Spearman correlation coefficients of pre-service teachers towards collaborative learning on the basis of cross-curricular relationships and problem-solving abilities

Intended educational practice and problem-solving activities	Intended educational practice Cross-curricular relationships will be applied	Collaborative learning will be applied	Management of information	Explanation of problems	Practical application of experience
Cross-curricular relationships will be applied	1.000	0.290(**)	0.122	0.095	0.227(*)
Collaborative learning will be applied		1.000	-0.083	0.053	-0.032
Management of information			1.000	0.069	-0.034
Explanation of problems				1.000	0.200(*)
Practical application of experience					1.000

**Correlation is significant at the 0.01 level (two-tailed)
*Correlation is significant at the 0.05 level (two-tailed)

Table 7 Problem-solving activities of learners

Steps of problem solving	Problems	Statements	Authors
Collaborative learning	Frustration	Experienced frustrations while working individually and in a group	Edwards and Hammer (2007)
	Lack of consensus	It was difficult to accept ideas or decisions of colleagues as less reasonable. Learners tended to refer to their own ideas	Barron, Preston-Sabin, and Kenedy (2013)
	Lack of experience	Learners lacked experience regarding problem-solving activities	Mohamed (2015)
	Insufficient harmonization of individual and group work	Learners experienced insufficient harmonization of an individual and group	Koh (2014)
Concretization of a problem	Evaluation of boundaries of a problem	Boundaries of a problem and its details were unclear. It was not clear how to start problem solving having no experience in a particular sphere	Edwards and Hammer (2007)
	Anticipation of the beginning of a problem	The beginning of a problem solution was difficult. Lack of knowledge in order to solve it	So, Yeung, Albert, and Volk (2001)
	The speed of defining a problem	Pre-service teachers were in a hurry defining problems without deeper analysis of problem-based situations	De Simone (2008)
Management of information	Hasty search of information	Independent search of information was difficult	McPhee (2002)
	Superficial analysis of the content of information	Learners experienced difficulties of choosing suitable sources of information	Yassin et al. (2011)
Prediction of problem-solving scenario	Wrong way of problem solving	Learners offered wrong ways of problem solving. They tried to guess the solution of a problem	De Simone (2008)
Monitoring of problem-solving process	Expectation of help during self-reflection	Low level of reflection and superficial expression of ideas	Yassin et al. (2011)

of problem-solving abilities corresponding to the second SPS strategy (implementation) ($\bar{x} = 71.54 \pm 17.32$) were higher than the means of problem-solving abilities of the first SPS strategy (identification) ($\bar{x} = 59.96 \pm 12.79$).

The study also revealed whether the means of the subscales differed statistically significantly from the point of view of problem-solving abilities (Table 5). ANOVA data block was used to identify the statistical significance among the means. This statistical criterion is applied for more than two dependent samples when the data are parametrical. Sphericity assumed (p = 0.000) showed the means that differed

Table 8 The results of the focus group interview: problem-solving cycle, categories and statements

Problem-solving cycle	Categories	Statements
Collaborative learning	Harmonization of educational activities of pre-service science teachers	Cross-curricular projects forced me to feel like a pupil studying Biology, Chemistry and Physics. For the first time in my life, I understood that a teacher of Physics should also think about the subjects that Biology and Chemistry teachers explain. It is necessary to know the programmes of other subjects and textbooks and the teaching styles as well (Asta)
		The implementation of a project is constantly related to thinking how to explain a phenomenon from the point of view of Physics, Chemistry and Biology. It is easier for me to analyse a problem from the point of view of Physics. The latter tasks are more understandable, and it is not necessary to go deep into them (Asta, Physics)
Concretization of a problem	Complex attitude towards cross-curricular problems	Performance of cross-curricular projects enabled me to understand that it was possible to discover many common links among Physics, Chemistry and Biology. It was possible to reduce the amount of information and make the studies easier (Rimas)
Management of information	Particularity of the search of cross-curricular information	Searching for subject information is quite an easy task. It is necessary to write the keyword, and you get what you want. Therefore, it is not suitable for implementing cross-curricular projects. There is a problem regarding keywords and cross-curricular relationships that should be found by ourselves. No other searching device can help you (Aušra)
Prediction of a problem-solving scenario	Harmonization of ideas while creating cross-curricular scenario of problem solving	Each of our projects is a small problem that is necessary to solve. Finding solutions to the problem is the most difficult part, while practice and technique play secondary roles. It is more difficult to decide what to do, not how to do (Rimantas)
Monitoring	Collaborative evaluation	Self-evaluation of cross-curricular projects is possible only while working in a group of students studying different subjects. It is necessary to discuss and help each other in order to achieve a common agreement on evaluation (Romas)

statistically significantly. Statistically significant differences of the means of the eighth ability (explanation of problem solution, see Table 4) and the other abilities were studied (Table 5).

The results revealed statistically significant differences of the mean of the eighth (Table 4) factor (explanation of problem solution) from other factors of the first cycle (identification of a problem). Such differences occurred due to the fact that the mean of the eighth factor (explanation of problem solution) was much higher than those of the first cycle of problem-solving abilities. The results of ANOVA disclosed statistically insignificant differences of the mean of the eighth factor in comparison to other factors of the second SPS strategy (implementation of problem solution).

We analysed the relationships between problem-solving abilities of the second problem-solving strategy based on cross-curricular contents and the attitude of pre-service teachers to use the cross-curricular relationships and collaborative learning in the future educational practice at school (Table 6). The analysis of the statistical relationships between second strategy problem-solving abilities and intentions of application of cross-curricular relationships and collaborative learning in educational practice was performed. Statistically significant correlation was revealed between the intention of usage of cross-curricular relationships and collaborative learning (C = 0.290**, p = 0.01). It means that the experience obtained by pre-service teachers during science education seminars would be applied in educational practice in the future.

5.2   Qualitative Research

The process of cross-curricular problem solving was analysed from the qualitative point of view using important activities of IBSL (collaborative learning, concretization of a problem, prediction of the problem-solving scenario, management of information and monitoring of problem-solving process). The review of scholarly literature (Barron, Preston-Sabin, & Kenedy, 2013; De Simone, 2008; Edwards & Hammer, 2007; Koh, 2014; Mohamed, 2015; Yassin, Rahman, & Yamat, 2011) revealed that collaborative problem solving enhanced the following negative aspects: frustration, lack of consensus, lack of experience as well as lack of individual coordination and collaborative activity (Table 7). A lot of challenges for collaborative problem-solving learning were also disclosed: learners worked and gathered information for the solution of problems at the superficial level, presented information without deeper analysis or its diversification, group work was not efficient, information was lacking and self-evaluation was quite slow (Yassin et al., 2011).

The analysis of literary sources revealed that some peculiarities were characteristic for the performance of nonintegrated projects: superficial analysis of information, limited usage of information resources and shortage of suitable questions (Table 7). Management of information was quite different during cross-curricular (or complex) problem solving (Fischer & Neubert, 2015; Funke, 2010; Sternberg & Frensch, 1991). The qualitative analysis of focus groups confirmed this issue. It was necessary to analyse different sources of information, as well as to perform reasonable search of information (e.g. "there is a problem regarding keywords and cross-curricular relationships that should be found by ourselves. No other searching device can help you", a quote from our research).

We used SPS strategies that were based on cross-curricular science content during the implementation of our study. The projects with cross-curricular content corresponded to the idea of inquiry-based strategy, as cross-curricular links among training subjects had a subjective character. Every learner could envisage specific cross-curricular relationships in the content of Biology, Chemistry and Physics. The results of our qualitative research revealed that the variety of links was disclosed better when the same project was implemented by a group of pre-service teachers studying at different (Biology, Chemistry and Physics) programmes. The variety of cross-curricular relationships highlighted important aspects of collaborative learning: an ability to achieve a common solution, agreement on one model of cross-curricular relationships and one scenario of problem solving (Table 8).

The qualitative analysis of focus groups revealed the main features of pre-service teachers' problem-solving abilities (Table 8). The participants of the research described peculiarities of collaborative problem solving and stressed coherence of educational activities of science teachers (e.g. "a teacher of Physics should also think about the subjects that Biology and Chemistry teachers explain", a *quote from our research*). The pre-service teachers experienced problems using individual knowledge in cross-curricular contents. We observed subject-related individualism and cross-curricular frustration (e.g. "it is easier for me to analyse a problem from the point of view of Physics", a *quote from our research*). The respondents mentioned that the phase of concretization of a problem (evaluation of the boundaries of a problem) on the basis of cross-curricular background was quite difficult, unspecified and complex (e.g. "the performance of cross-curricular projects enabled me to understand that it was possible to discover many common links among Physics, Chemistry and Biology", a *quote from our research*).

Self-evaluation of cross-curricular problem solving was, therefore, an intricate and complex task ("it is necessary to discuss and help each other in order to achieve a common agreement on evaluation"). Collaborative learning facilitated the self-evaluation of problem solving by pre-service teachers studying different science subjects. Individual self-evaluation of cross-curricular problem solving was not an easy task as participants lacked appropriate knowledge of other science subjects.

6   Discussion

IBSL in science education can be implemented by many scientific steps: refining investigation questions; formulating hypotheses; planning, managing and carrying out investigation; analysing and evaluating data; interpreting results; developing explanations; and communicating scientifically (Bybee, 2011). Inquiry-based learning (IBSL) is a framework for PBL and problem solving in science education. PBL activity belongs to open inquiry. In open inquiry learners should be able to organize and manage complex and extended activities. If learners lack these abilities, they will be unable to complete a meaningful investigation (Edelson, Gordin, & Pea, 1999).

We analysed problem-solving activities of pre-service teachers using the three stages of the problem-solving strategy: (1) identification, (2) solution and (3) checking. The three-stage approach facilitates the analysis of problem-solving process in education. The stages of problem-solving strategy were very useful, important and helpful at the open-ended activity to comprehend, solve and analyse cross-curricular problems.

The first stage – identification of a problem – is not easy. The results of our research show that the identification of cross-curricular problems is difficult for science students because it requires cross-curricular knowledge (Physics, Chemistry and Biology) and abilities. The results of our research show that pre-service science teachers self-evaluate the abilities of problem identification by a lower score than the abilities of problem implementation.

The cross-curricular contents create suitable conditions for the improvement of information management abilities, as it involves a relatively high number of steps, promotes the processes of dealing with situations involving many variables and helps in discovering undisclosed information. It means that the experience obtained by pre-service teachers during science education seminars will be applied in educational practice in the future.

The correlation between the abilities to use knowledge in practice and intention of using interdisciplinary relationships was also significant ($C = 0.227^*$, $p = 0.05$). The latter statistical relation could be explained that it was necessary to have abilities of knowledge application in different situations and contexts while using cross-curricular relationships in educational practice.

Collaborative problem solving helps learners to identify problem-solving strategies (Gillies, Nichols, Burgh, & Haynes, 2012). Learners work in mixed groups together in order to achieve common goals, and for this they have to discuss with each other and help each other to identify cross-curricular problems in cooperative open inquiry (Zsoldos–Marchis, 2014). Our research confirmed the ideas of Gillies et al. (2012) and Zsoldos-Marchis (2014) about the peculiarities of identifying a collaborative cross-curricular problem.

The cross-curricular problems were complex and required ingenuity in their solving. Through inquiry and investigations, learners realize that answers to problems do not readily appear, nor can they be found via quick reference to authority, but rather are solved through hard work and thinking (Trowbridge, Bybee, & Powell, 2004). Stepwise problem-solving strategies positively affect and facilitate the implementation of problem solving (Gok, 2014). Our results revealed that pre-service teachers self-evaluated the explanation of problem solution and practical application of experience, which was rated most favourably.

The solution of cross-curricular problems is quite a difficult task, as every problem is a unique one. Learners realize that answers to problems do not readily appear (Trowbridge et al., 2004). It is not easy to find necessary information using ordinary virtual search systems. The search of information changes while solving cross-curricular problems. The results of our research revealed that students were encouraged to analyse different sources of information, perform detailed and reasonable search of information, help each other to find suitable information and make it structural. It leads to the improvement of information management abilities.

7 Conclusions

1. Challenges of the twenty-first century require a return to the collaborative problem solving. CPS is a dynamic process unfolding over time, with more differentiation going beyond simple acquisition and application of abilities. The results of our research revealed that the means of problem-solving abilities of pre-service science teachers at different stages of SPS are different. The means of problem-solving abilities corresponding to the implementation stage were higher than the means of problem-solving abilities of the identification stage.

2. Collaborative problem-solving abilities of pre-service science teachers were related to the attitude of using the cross-curricular relationships and collaborative learning in the future educational practice at school. A statistically significant correlation was revealed between the intention of usage of cross-curricular relationships and collaborative learning.

3. The results of qualitative research about the role of cross-curricular content projects in terms of IBSL activity (collaborative learning, concretization of a problem, prediction of the problem-solving scenario, management of information and monitoring of problem-solving process) revealed that cross-curricular tasks encouraged pre-service science teachers to concretize the problem, predict a problem-solving scenario, manage information and monitor the problem-solving process.

4. The results of qualitative research showed that collaborative learning based on cross-curricular content encouraged the harmonization of educational activities of pre-service science teachers, formed a holistic view towards the formation of cross-curricular problems and adjustment of ideas while creating problem-solving scenarios and highlighted the role of collaboration during the evaluation of results of cross-curricular problem solving. We would encourage researchers in the field of CPS to come back to the idea of harmonization of educational activities and improving our understanding of how CPS deals with the idea of harmonization of educational activities.

Acknowledgements This work was supported by the EU Seventh Framework Programme project "Science-Teacher Education Advanced Methods (S-TEAM)" No. SIS-CT-2009-234870.

References

Banchi, H., & Bell, R. (2008). The many levels of inquiry. *Science and Children, 46*(2), 26–29.
Barron, L., Preston-Sabin, J., & Kennedy, D. (2013). Problem-based learning for the pre-service teacher. *Journal of the Southeastern Regional Association of Teacher Educators, 22*(2), 39–45.
Boud, D., & Feletti, G. (1997). *The challenge of problem based learning*. London: Kogan Page.
Brandsford, J. D., & Stein, B. S. (1984). *The ideal problem solver: A guide for improving thinking, learning and creativity*. New York: W.H. Freeman and Company.
Bybee, R. W. (2011). Scientific and engineering practices in K-12 classrooms: Understanding 'a framework for K-12 science education'. *Science Teacher, 78*(9), 34–40.

Capon, N., & Kuhn, D. (2004). What's so good about problem-based learning? *Cognition and Instruction, 22*, 61–79.

De Simone, C. (2008). Problem-based learning: A framework for prospective teachers' pedagogical problem solving. *Teacher Development, 12*(3), 179–191.

Dochy, F., Segers, M., Van den Bossche, P., & Gijbels, D. (2003). Effects of problem-based learning: A meta-analysis. *Learning and Instruction, 13*, 533–568.

Dörner, D., & Funke, J. (2017). Complex problem solving: What it is and what it is not. *Frontiers in Psychology, 8*, 1153.

Doymuş, K., Şimşek, U., & Bayrakçeken, S. (2004). The effect of cooperative learning on attitude and academic achievement in science lessons. *Journal of Turkish Science Education, 2*(2), 103–113.

Edelson, D. (2001). Learning-for-Use: A framework for the design of technology-supported inquiry activities. *Journal of Research in Science Teaching, 38*(3), 355–385.

Edelson, D. C., Gordin, D. N., & Pea, R. D. (1999). Addressing the challenges of inquiry-based learning through technology and curriculum design. *Journal of the learning sciences, 8*(3–4), 391–450.

Edwards, S., & Hammer, M. (2007). Problem based learning in early childhood and primary pre-service teacher education: Identifying the issues and examining the benefits. *Australian Journal of Teacher Education, 32*(2), 21–36.

Ferreira, M. M., & Trudel, A. R. (2012). The impact of problem-based learning (PBL) on student attitudes toward science, problem-solving skills, and sense of community in the classroom. *Journal of Classroom Interaction, 47*(1), 23–30.

Fischer, A., & Neubert, J. C. (2015). The multiple faces of complex problems: A model of problem solving competency and its implications for training and assessment. *Journal of Dynamic Decision Making, 1*(1), 1–14.

Florida, R., & Tinagli, I. (2004). *Europe in the creative age.* Pittsburg, CA: Carnegie Mellon Software Industry Center.

Foldevi, M. (1995). *Implementation and evaluation of problem-based learning in general practice.* Linköping University Medical Dissertations No. 473. Linköping, Sweden.

Funke, J. (1991). Solving complex problems: Exploration and control of complex systems. In R. Sternberg & P. A. Frensch (Eds.), *Complex problem solving: Principles and mechanisms* (pp. 185–222). Hillsdale, NJ: Lawrence Erlbaum Associates.

Funke, J. (2010). Complex problem solving: A case for complex cognition? *Cognitive Processes, 11*, 133–142.

Gallagher, S. A., Stepien, W. J., & Rosenthal, H. (1992). The effects of problem-based learning on problem solving. *Gifted Child Quarterly, 36*, 195–200.

General Programs of Elementary and Basic Education. (2008). *Vilnius: Švietimo aprūpinimo centras* (in Lithuanian).

Gillies, R. M., Nichols, K., Burgh, G., & Haynes, M. (2012). The effects of two strategic and meta-cognitive questioning approaches on children's explanatory behaviour, problem-solving, and learning during cooperative, inquiry-based science. *International Journal of Educational Research, 53*, 93–106.

Gok, T. (2014). Students' achievement, skill and confidence in using stepwise problem-solving strategies. *Eurasia Journal of Mathematics, Science & Technology Education, 10*(6), 617–624.

Hansson, A., Foldevi, M., & Mattsson, B. (2010). Medical students' attitudes toward collaboration between doctors and nurses – A comparison between two Swedish universities. *Journal of Interprofessional Care, 24*(3), 242–250.

Harlen, W., & Allende, J. (2006). *IAP Report of the Working Group on the International Collaboration in the Evaluation of IBSE programs.* Fundacion para Biomedicis Avanzados do la Facultad de Medicina, University of Santiago, Chile.

Hayes, J. R. (1989). *The complete problem solver* (2nd ed.). London: Routledge.

Hunt, E. (1994). Problem solving. In R. J. Sternberg (Ed.), *Thinking and problem solving* (pp. 215–232). San Diego, CA: Academic.

Johnson, D. W., & Johnson, R. T. (1999). Making cooperative learning work. *Theory into Practice, 38*(2), 67–73.

Jonassen, D. H. (2011). *Learning to solve problems: A handbook for designing problem-solving learning environments*. New York: Routledge.

Keebaugh, A., Darrow, L., Tan, D., & Jamerson, H. (2009). Scaffolding the science: Problem based strategies for teaching interdisciplinary undergraduate research methods. *International Journal of Teaching and Learning in Higher Education, 21*(1), 118–126.

Kirschner, P. A., Sweller, J., & Clark, R. E. (2006). Why minimal guidance during instruction does not work: An analysis of the failure of constructivist, discovery, problem-based, experiential, and inquiry-based teaching. *Educational Psychologist, 41*(2), 75–86.

Koh, K. (2014). Developing preservice teachers' assessment literacy: A problem-based learning approach. In P. Preciado Babb (Ed.), *Proceedings of the IDEAS: Rising to challenge conference* (pp. 113–120). Calgary, Canada: Werklund School Education, University of Calgary.

Kumar, M., & Natarajan, U. (2007). A problem-based learning model: Showcasing an educational paradigm shift. *Curriculum Journal, 18*(1), 89–102.

Linn, M. C., Davis, E. A., & Bell, P. (2004). *Internet environments for science education*. Mahwah, NJ: Lawrence Erlbaum Associates.

Mayer, R. (2004). Should there be a three-strikes rule against pure discovery learning? The case for guided methods of instruction. *American Psychologist, 59*(1), 14–19.

McPhee, A. (2002). Problem based learning in initial teacher education: Taking the agenda forward. *Journal of Educational Enquiry, 3*(1), 60–78.

Metallidou, P. (2009). Pre-service and in-service teachers' metacognitive knowledge about problem-solving strategies. *Teaching and Teacher Education, 25*, 76–82.

Mohamed, M. E. S. (2015). An investigation into pre-service teachers' perceptions of learning primary schools sciences using the method of problem based learning. *World Education Journal, 5*(3), 44–60.

Moskovitz, C., & Kellog, D. (2011). Science education inquiry-based writing in the laboratory course. *Science Education, 332*, 919–920.

OECD. (2013). *Education at a Glance 2013: OECD Indicators*. Paris: OECD Publishing. https://doi.org/10.1787/eag-2013-en

Oguz-Unver, A., & Arabacioglu, S. (2014). A comparison of inquiry-based learning (IBL), problem-based learning (PBL) and project-based learning (PJBL) in science education. *Academia Journal of Educational Research, 2*(7), 120–128.

Patton, M. Q. (2002). *Qualitative research & evaluation methods*. Thousand Oaks, CA: Sage Publications.

Pierce, W. (2008). Inquiry made easy. In E. Brunsell (Ed.), *Readings in science methods, K8* (pp. 33–36). Arlington, TX: NSTA Press.

Pierrakos, O., Anderson, R., & Barrella, E. (2016). *A developmental and adaptive Problem Based Learning (PBL) model across the curriculum: From theory to practice in integrating and assessing PBL experiences across the James Madison University engineering curriculum*. Erie, PA: 2016 IEEE Frontiers in Education Conference (FIE). https://doi.org/10.1109/FIE.2016.7757338

Pólya, G. (1945). *How to solve it*. Princeton, NJ: Princeton University Press.

Pretz, J. E., Naples, A. J., & Sternberg, R. J. (2003). Recognizing, defining, and representing problems. In *The psychology of problem solving* (pp. 3–30). New York: Cambridge University Press.

Reif, F., Larkin, J. H., & Brackett, G. C. (1976). Teaching general learning and problem-solving skills. *American Journal of Physics, 44*(3), 212–217.

Savery, J. R. (2006). Overview of problem-based learning: Definitions and distinctions. *Interdisciplinary Journal of Problem-Based Learning, 1*(1), 9–20. https://doi.org/10.7771/1541-5015.1002

Savery, J. R. (2015). Overview of problem-based learning: Definitions and distinctions. In A. Walker, H. Leary, C. Hmelo-Silver, & P. A. Ertmer (Eds.), *Essential readings in problem-*

based learning: Exploring and extending the legacy of Howard S. Barrows (pp. 5–15). West Lafayette, Indiana: Purdue University Press.

Savin-Baden, M. (2005). Learning spaces, learning bridges and troublesomeness: The power of differentiated approaches to problem-based learning. *Problem-Based Learning: New Directions and Approaches, 1*(1), 10–28.

Savin-Baden, M. (2016). The impact of transdisciplinary threshold concepts on student engagement in problem-based learning: A conceptual synthesis. *Interdisciplinary Journal of Problem-Based Learning, 10*(2). https://doi.org/10.7771/1541-5015.1588

Schmidt, H. G., Van der Molen, H. T., Te Winkel, W. W., & Wijnen, W. H. (2009). Constructivist, problem-based learning does work: A meta-analysis of curricular comparisons involving a single medical school. *Educational Psychologist, 44*(4), 227–249.

So, K. S., Yeung, K. H., Albert, T. K., & Volk, K. (2001). Introducing problem-based learning to teacher education Programmes. In D. Kember, S. Candlin, & L. Yan (Eds.), *Further case studies of improving teaching and learning from the action learning project.* Hong Kong, China: Action Learning Project.

Sternberg, R. J., & Frensch, P. A. (1991). *Complex problem solving: Principles and mechanisms.* Hillsdale, NJ: Erlbaum.

Trowbridge, L. W., Bybee, R. W., & Powell, J. C. (2004). *Inquiry and conceptual change. Teaching secondary school science: Strategies for developing scientific literacy.* Upper Saddle River, NJ: Pearson Education.

Welch, W. W., Klopfer, L. E., Aikenhead, G. E., & Robinson, J. T. (1981). The role of inquiry in science education: Analysis and recommendations. *Science Education, 65,* 33–50.

Yassin, S. F. M., Rahman, S., & Yamat, H. (2011). ICT interdisciplinary problem-based learning in pre-service teacher Programme. *World Applied Sciences Journal (Innovation and Pedagogy for Lifelong Learning), 15,* 42–48.

Zoller, U. (2011). Science and technology education in the STES context in primary schools: What should it take? *Journal of Science Education and Technology, 20*(5), 444–453.

Zsoldos-Marchis, I. (2014). Influence of cooperative problem solving on students' control and help-seeking strategies during mathematical problem solving. *Acta Didactica Napocensia, 7*(3), 49–59.

Teachers as Educational Innovators in Inquiry-Based Science Teaching and Learning

Anni Loukomies, Kalle Juuti, and Jari Lavonen

1 Introduction

This chapter introduces a professional development programme (PDP) for science teachers. The PDP was built on the principle of acknowledging teachers' expertise in designing their teaching. The participants were provided with a theoretical grounding in inquiry-based science teaching and learning (IBST/L) and its benefits for students' motivation, interest and learning science—in other words, a rationale for employing the IBST/L approach. The PDP sought to support teachers in the process of collaboratively reflecting on their existing practices and revising them with a view to employing the principles of IBST/L in teaching. In what follows, the research-based understanding of teachers' expertise, the meaning of IBST/L and its potential to promote pupils' and students' motivation, interest and learning are introduced. As what follows, the application of these IBST/L ideas in a PDP that employed reflection activities as a means of recognising and revising the teaching practices of participating teachers is described. Then the results of the data analysis are presented. The data consists of teachers' poster presentations of their IBST/L

A. Loukomies (✉)
Faculty of Educational Sciences, Viikki Teacher Training School, University of Helsinki, Helsinki, Finland

Department of Childhood Education, University of Johannesburg, Johannesburg, South Africa

K. Juuti
Faculty of Educational Sciences, Department of Education, University of Helsinki, Helsinki, Finland

J. Lavonen
Faculty of Educational Sciences, Department of Education, University of Helsinki, Helsinki, Finland

Department of Childhood Education, University of Johannesburg, Johannesburg, South Africa

© Springer International Publishing AG, part of Springer Nature 2018
O. E. Tsivitanidou et al. (eds.), *Professional Development for Inquiry-Based Science Teaching and Learning*, Contributions from Science Education Research 5, https://doi.org/10.1007/978-3-319-91406-0_10

pilots and video-recorded reflection sessions during the PDP. Finally, the potential and challenges of organising a PDP based on such an approach are discussed and, more specifically, how this approach served to promote IBST/L in the participants' teaching practices.

1.1 Inquiry-Based Teaching and Learning (IBST/L) and Professional Development

The aims of school science encompass more than learning science concepts or learning to perform scientific experiments. It is essential, of course, to understand concepts and the relations between them, but the scope and the aims of learning and teaching science are more comprehensive, encompassing scientific thinking, a coherent worldview and students' development as learners. As elsewhere, this is the situation in Finland. Along with the understanding of scientific concepts, the Finnish national science curriculum that takes effect from 2016 (FNBE, 2014) introduces the following aims, among others, for science learning: establish the motivation to learn science and the ability to take responsibility for one's learning; set aims for learning; think critically and in a scientific way; plan, conduct and communicate scientific investigations in a collaborative manner; and use models and concepts to explain scientific phenomena. The new science curriculum also emphasises practices that are important in engineering, such as the use of creativity in science- and technology-related projects. At a broader level, students are expected to develop a coherent and scientifically argued view of the world, as well as an understanding of how scientific knowledge is generated. They are also expected to act responsibly in relation to their environment and to make reasonable decisions based on scientific thinking.

Beyond Finland, the active renewal of science curricula is in evidence across the world. For example, in the USA, the Next Generation Science Standards (NGSS) that frame K–12 science education also aspire to students' comprehensive understanding. These standards introduce the idea of science practices that encompass a variety of skills and knowledge related to the actual behaviours of scientists. In particular, the standards emphasise that science practices should not be reduced to traditional experimental investigations but should also encompass other significant aspects of science, such as evaluation, evidence-based argumentation and communication.

In recent science education research, aims and characteristics of the kind described above are typically referred to as *inquiry-based*. However, it has also been argued that this term has been interpreted in too many different ways by the science education community in the context of science education (NRC, 2012, p. 30). Given this confusion about what is specifically meant by inquiry in the first place, we have adopted the framework constructed by Minner, Levy, and Century (2010), both for its soundness and because it emphasises issues considered central in the Finnish National Core Curriculum for Science Education.

Minner et al. (2010) constructed this framework following their review of 138 studies of inquiry instruction, from which they extracted the following 6 common characteristics: (1) presence of science content, (2) student engagement with science content, (3) student responsibility for learning, (4) active thinking by students, (5) student motivation and (6) an investigation cycle that encompasses formulating the question to be investigated, designing the investigation, collecting and organising data, drawing conclusions and communicating the investigations. Items 3, 4 and 5 should occur within at least one of the components of the investigation cycle. These essential process skills are associated with the scientific method and understanding of 'the nature of science' that Anderson (2007) includes in his definition of inquiry. Anderson also specifies that student engagement with science content encompasses epistemologically authentic procedures such as reasoning, asking questions and designing experiments, and he further emphasises the role of social interaction and collaboration.

The definition of inquiry described above, which is based on the work of Minner et al. (2010) and Anderson (2007), is what was passed on to the teachers who were participating the PDP. It is in line with the one introduced in the Chapter "What Is Inquiry-Based Science Teaching and Learning?" of this volume. In Chapter "What Is Inquiry-Based Science Teaching and Learning?" it is emphasised that 'inquiry involves a degree of autonomy or responsibility for learning' (p. 9, this volume). When organising the PDP, the fundamental principle was to respect the participant teachers' autonomy during their planning and designing process and encourage them to further offer the same autonomy to their pupils. In Chapter "What Is Inquiry-Based Science Teaching and Learning?" and in the definition above, also other aspects of inquiry are emphasised besides those related to motivation, responsibility and active thinking, but in this chapter, the focus is on these three aspects. Understanding the nature of science and the scientific processes may be in the focus in future research.

Promoting motivation is important because there is evidence that, whatever one's theoretical perspective, high-quality motivation yields better learning outcomes (Guay, Ratelle, & Chanal, 2008; Niemiec & Ryan, 2009; Reeve & Halusic, 2009) and better mental well-being (Tuominen-Soini, 2012; Vasalampi, Salmela-Aro, & Nurmi, 2009). Teachers play a key role in supporting students' motivation and interest, and therefore the basic mechanisms of generating and maintaining motivation in the contexts of the self-determination theory and expectancy-value theory (Eccles, 2005; Ryan & Deci, 2002) were introduced to the participants.

In Finland, teachers are valued as experts in curriculum development, teaching and assessment at all school levels. At the same time, however, Finnish science teachers tend to be pedagogically conservative, commonly favouring direct teaching of large groups of students (Juuti, Lavonen, Aksela, & Meisalo, 2009; Norris, Asplund, MacDonald, Schostak, & Zamorski, 1996). It is often quite difficult to convince teachers to change their teaching practices, and this is one of the crucial issues in teachers' professional development (e.g. Donovan, Bransford, & Pellegrino, 1999). However, Yeager and Walton (2011) suggest in their review that small-scale interventions that are firmly grounded on relevant theories but also take into account the characteristics of the context in which they are about to take place may have large-scale and lasting effects. They stressed that the focus of these interventions is not on learning the content but changing the participants' mindsets through using persuasive methods that are grounded on relevant research and getting the participants to communicate the new ideas.

1.2   Teachers as Educational Innovators and Reflective Professionals

Finnish teachers are considered to be professionals with high-level subject knowledge, pedagogical content knowledge and general pedagogical knowledge. Teachers are expected to have a good understanding of student assessment and curriculum development. As professionals, teachers must also exhibit high-level communication skills and moral knowledge, as well as the skills needed for professional development (Krzywacki, Lavonen, & Juuti, 2015). Instead of reading and following detailed descriptions of lesson plans, teachers should design their own applications of newly introduced pedagogical approaches. The vision of teachers from Keith Sawyer (2004) is adopted in this research. Sawyer insisted that '[T]eachers are knowledgeable and expert professionals and are granted creative autonomy to improvise in their classroom' (p. 18). To complement this view, Lavonen, Juuti, Aksela, and Meisalo (2006) identified empowerment and communication as 'optimal features' for the professional development projects of Finnish science teachers. By *empowerment* they meant that, in their professional development practices, teacher educators must consider teachers as professionals who make independent decisions about their own teaching. It follows that 'teachers should be guided in their planning and evaluation of small teaching experiments that are then implemented in their schools together with the assistance of other teachers' (Lavonen et al., 2006, p. 170). *Communication* aspects were seen to include optimal pace and creative atmosphere.

Taken together, these two 'optimal features' emphasise that PDPs cannot be designed as 'scripts' for teachers to follow. The combination of lectures and formal and informal small-group activities in a PDP must be sufficiently flexible to enable teacher educators to accommodate participant teachers' reactions, with time allocated for the development of ideas. As argued by Penuel, Fishman, Yamaguchi, and Gallagher (2007), offering participants time and support to plan their interventions is important for integrating the new course content into their teaching practice. Lavonen et al. (2006) emphasised that teacher professional development projects must allow room for the free generation of ideas and for positive feedback on all ideas. The atmosphere must be safe and supportive to encourage teachers to take risks despite the possible 'failure' of teaching experiments. Juuti et al. (2009) highlighted the importance of providing opportunities for informal communication in a supportive atmosphere.

According to Sawyer (2004), a scripted curriculum fails to access either teachers' creativity or their subject knowledge. Introducing the metaphor 'teaching as improvisation', Sawyer makes an analogy between classroom discourse and improvisational theatre, where actors work without scripts. Instead, they have only broad structures or 'games' that they play. Sawyer summarises sociocultural and social constructivist theory as implying, in a sense, that effective teaching must be improvisational; otherwise, students cannot co-construct their own knowledge. A classroom where the teacher controls the discussion is not improvisational. Rather, improvisational classrooms are collaborative, drawing on constructivist and inquiry-based methods. 'In improvising, the teacher creates a dialogue with the students, giving them freedom to creatively construct their own knowledge' (Sawyer, 2004, p. 14). To support this construction process, teachers need to offer disciplinary tools for students to refine their thinking about science content (Jurow & McFadden, 2011).

In helping teachers to develop their teaching practices, teaching experiences need to be reflected upon. Reflective thinking is an essential aspect of teachers' professional development, as it connects beliefs and practice (Mansvelder-Longayroux, Beijaard, & Verloop, 2007). Through reflection on actions, an experience becomes knowledge, and collaborative reflection with peer teachers and researchers creates new knowledge. As belief change follows changes in practice (reflection on action), teachers should be guided to monitor their practice and to reflect their new experiences. Reflection and learning from different perspectives can be facilitated through sharing within collaborative discussions. To encourage teachers to integrate educational innovation and research into their practices and beliefs, it is important to organise activities that can support collaboration and reflection (Uhrich, 2009). According to Rodgers (2002), reflection depends on attitudes that value personal and intellectual growth, both in oneself and in others (p. 845). The PDP approach described here sustains such attitudes by appreciating teachers as professionals who design their own work rather than as passive recipients of knowledge.

1.3 Research Questions

When designing the workshops, the aim was to promote participating teachers' professional development and growth. The aim was also to facilitate teachers' adoption of theory-based aspects of inquiry, supported by an understanding of its benefits for students' motivation and learning. Finally, the aim was to help them to adopt features of IBST/L in their own teaching, focusing on teaching strategies rather than on science content. In planning, the following aspects were taken into account: (1) Participating teachers are expected to be experts who autonomously design and develop their own teaching. Teachers and researchers appreciate each other's expertise; while teachers are experts on praxis (involving, e.g. knowledge of their students and groups), researchers are experts on theory. (2) In the contact meetings, researchers made brief introductory presentations on the principles and other aspects of IBST/L. Associated collaborative discussions assisted understanding of how the ideas presented would be of benefit in participants' classes. (3) The IBST/L pilots were planned to be designed and implemented by teachers in their classes independently and then presented and reflected in a PDP group session.

The focus of this research is not on the students' achievements or the development of their motivation but on how the participating teachers reported having implemented the ideas presented during the course. The implementations are evaluated with respect to the IBST/L frame suggested by Minner et al. (2010). The research questions are:

1. What were the teachers' inquiry-based science teaching and learning (IBST/L) pilots like?
2. How do the IBST/L aspects suggested by Minner et al. (2010) (support to students' active thinking, students' responsibility and students' motivation) appear in the teachers' pilots according to what the teachers presented?

2   Method

2.1   Participants and Outline of the Professional Development Programme (PDP)

In Finland, teachers have access to a wide variety of professional development courses, but participation in professional development is not compulsory. There was an open call to participate this PDP. Six lower secondary school physics and chemistry teachers participated in the PDP. All of them had major in either physics or chemistry and several years of teaching experience. All the participants were female, ranging in age from about 30 to 60 years.

Participating teachers could choose the scientific content they wished to work with from the school curriculum, and they were then asked to apply the principles of IBST/L in teaching the chosen content. Participants were introduced with the basic ideas of inquiry, the theoretical connection between inquiry teaching and motivation and the diverse possibilities for applying IBST/L. To maximise opportunities for collaborative discussion, the PDP was designed to be short and intensive (Sawyer, 2004, 2006; Yeager & Walton, 2011). The PDP was therefore run as two 2-day workshops. Between the workshops, each teacher implemented their IBST/L pilot in their own classroom.

During the first workshop, teachers planned their own pilots, which were then discussed collaboratively. It was agreed that instruction should follow the principles of IBST/L, with special emphasis to be placed on the engagement phase. The pilots were to consist of one or two 45–75-min lessons. Planning activity took place in pairs, but there was also a lot of ongoing collaborative discussion between pairs. During the workshop, the basic ideas of IBST/L were collaboratively discussed to encourage participants to reflect on their pilots. Each participant made a brief presentation on their own IBST/L pilot, followed by collaborative reflection. As short-term workshop-based programmes have been criticised for not being effective (Lumpe, 2007), it was decided to compensate for the brevity of the PDP intervention by emphasising effective feedback, cooperation, collegiality, practice-oriented staff development and a culture of shared beliefs and relationships (cf. Kim, Lavonen, Juuti, Holbrook, & Rannikmäe, 2013; Lumpe, 2007).

After the first workshop, the teachers went back to their schools and implemented their plans. In the second workshop, the teachers presented the implementations of their pilots, one after another. After each presentation, the implementation of the pilot was discussed. This workshop took 4 h, and it was video-recorded.

2.2   Data Collection and Analysis

The research questions are answered based on data that consists of teachers' posters of their pilots and a 4-h collaborative teacher presentation and reflective discussion that was video-recorded. The analysis of the data followed principles of theory-driven content analysis (Patton, 2002). The analysis categories in the analysis frame deductively emerged from theories, in more detail from the conceptualisation of IBST/L constructed by Minner et al. (2010), and theories conceptualising the components of motivation, in more detail the expectancy-value theory (e.g. Eccles, 2005) and the self-determination theory (e.g. Ryan & Deci, 2002). According to the frame suggested by Minner et al. (2010), an instruction was classified as inquiry if at least one of the phases of instruction included student responsibility, student active thinking and support for motivation. In our version, one minor revision was made with respect to the motivation subcategories. Minner et al.'s indices of motivation included students' expressions of interest, involvement, curiosity, enthusiasm, perseverance, eagerness, focus, concentration and pride. Many of these factors are closely related to intrinsic forms of student motivation and are therefore difficult to track for data such as poster presentation. For that reason, we enriched the approach by drawing on the motivation components introduced in the expectancy-value theory (Eccles, 2005) and the self-determination theory (Ryan & Deci, 2002).

Aspects related to the mentioned theories were extracted from the pilot posters and from those parts of the reflection discussion that concerned a certain pilot. These aspects were then categorised into the categories student responsibility, student active thinking and support for motivation. The motivation category consisted of subcategories: intrinsic value/interest (IN), attainment value/significance (AT), utility value (UT) (Eccles, 2005), support for autonomy (AU), support for competence (CO) and support for social relatedness (SR) (Ryan & Deci, 2002).

Each pilot was analysed using the template introduced in Table 1. In Table 1, the analysis of the pilot 1 is presented as an example of how the analysis was conducted. After the extraction the IBST/L aspects of each pilot and categorisation of them, a short description was written of each pilot. Then a more detailed description was written of the IBST/L features related to motivation, active thinking and responsi-

Table 1 Analysis tool for the IBST/L pilots

Teacher No. 1	
Science content	Physics, simple machines
Type of Engagement	Using Lego characters, students were required to construct a comic strip related to the topic *Simple machines*. They designed the storyline, constructed the scene and took the photos needed for the comic strip. Students also built the machines they needed to meet the physics content required by the instruction. The instruction was published in a web-based learning environment. At the end of the process, students commented each other's work.

Motivation
  Autonomy (AU)
  Competence (CO)
  Social Relatedness (SR)
  Intrinsic value/interest (IN)
  Attainment value/
  significance (AT)
  Utility value (UT)
	AU, SR, IN: Working on the comic strip
	AU, SR: Building the machines needed for the comic strip
	AU: Deciding on the length of the comics
	SR, AU: Choosing the groups
	AU, SR: Setting the scene
	AU, SR: Conducting the experiments autonomously
	AU: Deciding on the working order
	AU: Deciding on the story
	AU: Deciding on the software
	AU: Deciding on the allocation of tasks
	IN: Drawing a comic strip
	IN: Using Lego characters
	IN: Using language that fits the students' world
	AU: Working on a web-based learning environment
	IN, AU: Working with students' own cameras
	UT: Working on a web-based learning environment
	CO, AU: Finalising output autonomously
	CO: Using relevant physics concepts
	CO: Providing feedback for others in the web-based learning environment
	CO: Commenting on others' work
	UT: Using the web-based learning environment
	UT: Studying a topic from the curriculum

Responsibility	Working on the comic strip Working in a web-based learning environment Planning the comic strip Deciding on roles within the group Choosing the groups Deciding on task allocation Deciding on software Finalising the comic strip Providing feedback to other groups in the web-based learning environment
Active thinking	Interpreting the instructions provided in the web-based learning environment Recalling one's own experiences of simple machines Working on the comic strip Working in a web-based learning environment Building the machines Broadening the context beyond physics and creating a story Setting the scene and conducting experiments autonomously Deciding on the working order Using relevant physics concepts

bility. The phase of the procedure in which the IBST/L features occurred was noted, and the course of the lesson was compared to the phases of inquiry instruction proposed by Minner et al. (2010) (formulating the question to be investigated, designing the investigation, collecting and organising data, drawing conclusions and communicating the results of the investigation).

The content of each pilot (based on the poster and teacher presentation) is summarised below, along with a short description of its inquiry aspects. As this chapter focuses on the inquiry aspects of the pilots, their effect on students' motivation or learning is not evaluated here. It must be emphasised that the researchers were not present when the teachers implemented their pilots, but the analysis is conducted based on what the teachers reported about their implementations. For the purposes of the study, only five teachers' ISBT/L pilots were analysed; one teacher's plan was omitted from the analysis because of practical problems on the school side during implementation.

3   Results

3.1   Teacher 1

The context of this pilot was mechanics or, more precisely, simple machines. The pilot opened with a framework story, set in a world of Lego policemen and thieves. Students designed the scene for the story and photographed it. They had to consider how the characters' tasks were facilitated by certain simple machines, and they then had to translate the facts into explicit mathematical forms. The students worked in groups; the teacher assigned specific roles to all group members. The required output was a comic strip, constructed from their own photos of the scenes and continuing the story introduced by the teacher. The essential principles of simple machines were to appear in the comic, and these outputs were to be uploaded to a web-based learning environment for further collaborative elaboration. The comic strips served as a substitute for the traditional lab report. The teacher reported that some of the students did better than usual, and she was astonished at the students' specific use of physics concepts.

Students' Responsibility for Learning Teacher #1 allowed students to make decisions about groups, task allocation within the groups, the software to be used, the storyline behind the comic strip and how the scene for the story would be built. Students built the machines they needed to match the required physics content. They worked autonomously in groups according to the instructions published in the web-based learning environment. They also commented on each other's work.

Students' Active Thinking Teacher #1 required the students to interpret the web-based instructions independently. Students also needed to plan the scenes of their comics and to consider how the physics content was to be presented. They also needed to decide which physics concepts were needed.

Students' Motivation As well as motivation-related aspects emphasising students' responsibility, their feelings of social relatedness were supported by allowing them to work in groups. This teacher chose a framework story with Lego characters that contained intrinsic value. According to the teacher, the students also managed the task well and so increased their feelings of competence. Working in a web-based learning environment has utility value, as students learn skills that may be of subsequent use. However, there was little support for students' attainment value—the feeling that they are doing something significant—as the topic was not considered from any wider perspective beyond the classroom. *Simple machines* is also a topic on the primary school science curriculum in Finland, and lower secondary students might, for instance, have been asked to construct animations to be used in primary science teaching.

Nevertheless, of the four pilots designed during the PDP, this was the most multi-faceted and allowed most space for the students' creativity while at the same time highlighting disciplinary learning. The structure of this pilot was loosely compatible with the procedure proposed by Minner et al. (2010), as it encompassed the students' own design and encouraged them to implement the design and communicate the results. The major deficiencies were that the project did not begin with a research question and no experiments were conducted.

3.2   Teachers 2 and 3

These two teachers worked at the same school and therefore decided to plan a teaching sequence together. Only one teacher presented their pilot, which again involved the use of comic strips. The content sequences chosen by both teachers concerned nutrients. At the beginning of the pilot, the teachers first told the students about some of their own experiences related to nutrients, explaining how they taught nutrition, how they implemented this knowledge in their everyday lives and how they intended to combine the science content with the students' everyday experiences. They also discussed the typical structure of a three-panel comic strip and showed some examples. Students read the content of their own textbooks, and the teacher gave instructions how to transform the text into a comic strip. The teachers also gave instructions about the working schedule, group formation and specific tasks within the group. The students decided on the topic they wished to deal with and then allocated tasks within the group. The comics were constructed and the work was evaluated.

Student Responsibility Teachers chose the groups but allowed the students to allocate tasks within the group. Students were also allowed to choose their perspective within the limits of the topic of the lesson. Teacher #2 said it was usual for her to tell students about her own experiences, and in that sense at least, this lesson was no exception. In mid-process, the students modified the instruction, and the teacher allowed that to happen, as the students were eager to discuss the topic.

Active Thinking Students needed to consider how they would convert the text into a comic strip and to decide on the essential aspects of the text to be presented in a simple series of pictures. Students also had their own roles within the group, and they needed to decide what tasks were related to each role.

Motivation Students' feelings of social relatedness were supported by allowing them to work in groups. The drawing task was interesting for the most of the students and could be said to contain intrinsic value. It can also be said that the topic (Nutrients) contains some utility value, as it may benefit students' health in later life to have accurate information about the topic. However, from the IBST/L perspective, the structure of this lesson seemed quite traditional, and it was not compatible with the IBST/L procedure proposed by Minner et al. (2010). There was no actual research question to begin the procedure, nor did it begin from students' questions. Additionally, the lesson did not include any experimental work or data collection of any kind. The teachers made all the main decisions, as is readily apparent from the pilot poster, in which almost all sentences begin with phrases such as 'The teacher described the topic…', 'The teacher divided the students into groups…' and 'The teacher assigned topics to the groups…'. Clearly, not many decisions were left to the students, and there was no attempt to emphasise the meaningfulness of the task to increase attainment value.

3.3   Teacher 4

The scientific content of this pilot related to magnetism. The teacher decided to combine fictional and personal stories as well as concept maps and essay writing in her pilot sequence. Opening with a video of Superman and a magnetic telescope, she described her own experience of visiting a lab with a powerful magnet. Within the groups, students discussed what was fact and what was fiction in the film. According to the teacher, this was a very difficult task for the students. The experimental work relating to magnetism was conducted according to online video instructions. Students elaborated further on the topic by constructing concept maps, using *CmapTools* software. Their concept maps were based on the textbook, supported by facilitating questions. As homework, the students wrote essays based on their concept maps.

Student Responsibility Teacher 4 said that her students were not allowed to decide about anything, but they were expected to autonomously regulate their activities in constructing concept maps and writing essays. Their responsibility was emphasised in the essay-writing phase, and to prevent them from copying and pasting, the teacher asked them to write the essays with paper and pen.

Active Thinking In producing their concept maps and essays, students had to actively process their knowledge constructions. They also needed to consider how the web-based inquiry instruction was to be interpreted.

Motivation Teacher 4 included many interest-awakening situational features in the pilot, as students watched videos and heard stories about interesting occasions. This teacher found it very easy to tell spontaneous stories but had more difficulty telling something that had been decided beforehand. She included stories of many kinds in her pilot—a fictional story on video and her own experiences of the same topic. However, when constructing the poster about her pilot, she framed *motivation* as an isolated part of the lesson; in fact, what she thought of as motivation was merely awakening situational interest.

Although this pilot included a range of teaching methods and encompassed experimental work and although the students were required to work autonomously according to the teacher's instruction, the approach seemed quite traditional. After awakening interest, the teacher directed students to complete experimental work according to the instructions before constructing a concept map and writing an essay. The topic's connection to real-world problems was not emphasised, and the students were not allowed to decide their own perspective or to construct their own questions.

3.4   Teacher 5

The scientific content of this pilot related to nutrients. The lesson opened with a discussion of the students' own experiences of carbohydrates. After that, students collaborated in constructing concept maps with the *CmapTools* software, with each group outputting one shared concept map. Experimental work related to the same topic was also included in the lesson. As a homework output, students wrote brief reports on the experiment.

Student Responsibility Teacher 5 complained that she encountered a lot of practical difficulties in implementing her pilot. She had originally planned to organise the groups herself but then decided to leave this to the students themselves. The topic and experimental task were decided by the teacher, but other decisions were left to the students, such as the substances they chose to investigate. The students worked autonomously to produce reports on their experiments, and they constructed concept maps in groups. The teacher also encouraged students to describe their own experiences.

Active Thinking Students needed to actively process their knowledge constructions when producing concept maps, and they were active in drawing conclusions.

Motivation As well as being given responsibility, students worked with a topic that the teacher believed would be of interest to them. As knowledge of nutrients may be beneficial in later life and outside school, it can be said that the pilot had some utility value. The students also worked in groups; according to the SDT, this should promote their feelings of social relatedness and support their motivation.

Nevertheless, referring again to the essential features of IBST/L, this pilot can be said to have followed a traditional structure rather than the structure of an IBST/L lesson as proposed by Minner et al. (2010). From the beginning of the lesson, the teacher was pulling the strings, and no space was left for the students' own questions or fields of interest. The topic was not connected to the students' own lives, and so the attainment and utility value of the activities was not emphasised. In terms of Minner et al.'s framework, the nutritional content of food and issues such as sugar concentration may have invited more open investigations. For example, the students could have examined what they had eaten during a certain period and then reviewed the composition of their diet.

In summary, the four designed pilots can be described as quite traditional and teacher-centred. The teachers took responsibility for offering students adequate support and structure for their work, but a shared characteristic of these pilots was that little was left for students to decide. None of the pilots started with the students' questions; all were related to scientific content, but the teachers emphasised concepts. All the pilots contained some physical activities, in the form of experimental work, concept maps, comic strips or PowerPoint presentations. The motivational potential of these pilots related mainly to supporting students' autonomy by allowing them to decide something that was related to their work, supporting students' social relatedness by allowing them to work in pairs or groups and supporting intrinsic value (interest and enjoyment) mainly in terms of situational interest. Utility value was supported by choosing important topics from the curriculum but not from the point of view of the students' own lives and environment. The feeling of competence was not systematically supported—for example, there was no procedure for evaluating the work and giving feedback to peers or evaluating one's own work. Also, missing from all the pilots was any true support for attainment value— the feeling of doing something significant related to students' own lives or environment. With more support for attainment value, students' feeling of participation could also have been increased.

4   Discussion

In Chapter "What Is Inquiry-Based Science Teaching and Learning?" of this volume, a question was raised: How to encourage sufficient and effective inquiry-based science teaching of good quality? In this piece of research, this question was approached from two perspectives. First the idea was welcomed that a relevant way to promote change in science classes is to take the teachers' autonomy with respect to their teaching as a starting point. The central inquiry principle, namely, fostering autonomy and responsibility, was chosen as an approach when the teachers were guided in their planning process. Second, the PDP was strongly theory based. This principle was realised at two levels. The PDP itself was designed according to research-based principles of IBST/L and motivation. Furthermore, the content of the PDP was also based on research-based knowledge on motivation and student engagement. To sum up, the aim was to promote the effectiveness of the PDP by fostering the participant teachers' autonomy. Further, the quality of the teachers' IBST/L pilots was promoted through offering them a sound theoretical basis on how to support their students' engagement.

The teachers' pilots reflected their understanding of the IBST/L approach as introduced during the contact meetings. The pilots varied according to how teachers interpreted the content of the course. This variety of outcome strengthens the argument of Penuel et al. (2007) that teachers' interpretations of PD activities are important in shaping the effectiveness of those activities, beyond the design of the activities themselves. Each teacher interpreted the task guidelines in a way that fit their own ideas about how best to apply the principles and practices of inquiry teaching. The student groups were all different, which strengthened the improvisational aspect of teaching, even though the lessons were planned beforehand. Sawyer (2004) refers to this effect in arguing that teaching conceived as improvisation emphasises the teacher's creativity in responding to a unique group of students and the unpredictable flow of classroom discussion.

Based on the analysis, it can be suggested that the pilots contained some features of the IBST/L framework proposed by Minner et al. (2010). For instance, all the teachers allowed the students to work in groups, supporting collaboration. Teachers also expected the students to work autonomously and in a responsible way, and they enriched the pilots with material that would engage students' interest. However, the structure of the pilots remained essentially traditional. None of the teachers began the process from the students' own questions, nor did they expand their perspective beyond the classroom or encourage the students to do something that would have been truly significant at some level.

What, then, might account for the lack of inquiry features in these pilots? It is known that teachers' professional development tends to be very slow (Kim et al., 2013; Nelson, 2009). After reviewing several studies of the effects of short-term interventions, Laursen, Liston, Thiry, and Graf (2007) argued that, despite the popularity of the short-term intervention model, there is little convincing or statistically significant evidence of its effectiveness (p. 50). They suggested, however, that participants usually enjoy these occasions. On the other hand, Yeager and Walton (2011) suggest that even brief, the intervention may be effective because it sets into motion recursive social, psychological and intellectual processes in school level and within the individual (Yeager & Walton, 2011, p. 286).

Certainly, keeping PD events short is more likely to attract Finnish teachers, as they do not usually have the time, the interest or the financial resources to participate in more extended programmes (Taajamo, Puhakka, & Välijärvi, 2014), even though they are expected to acquire knowledge of many topics and novel teaching approaches. However, while a shorter PDP cannot change the way teachers teach, our vision of teachers as knowledgeable and expert professionals who are the agents of their own professional development suggests that even a short programme may help to shift their perspective.

In general, teachers tend to view inquiry-based science teaching as laborious, time-consuming and therefore difficult to apply in the classroom (Bybee, 2000).

The central principle of the PDP was not to tell teachers how to teach. Instead of asking participants to undertake any extracurricular pilots, they were asked to use the ideas from contact meetings in their teaching, to whatever extent they believed was appropriate for their own classes. The participants were shown that IBST/L is a perspective on teaching and learning rather than a highly structured teaching method, and it involves students' own responsibility for learning, active thinking and motivation during any lesson (cf. Minner et al., 2010).

During the workshops, the participating teachers' motivation was boosted by supporting their psychological needs (Ryan & Deci, 2002). The teachers were convinced that they were knowledgeable enough to design IBST/L pilots and that nobody else could in fact design their teaching (so increasing their feelings of competence). It was also communicated that teachers would be given a lot of autonomy to design their own pilots (feelings of autonomy) and that experiences were shared in a group (feelings of relatedness). Further, in preparing their pilots, the teachers were encouraged to choose a topic that interested them, to design something they could really use (utility value) and that might be of benefit to their students' learning and motivation (attainment value).

To foster the IBST/L approach, it may have been beneficial to include more support for planning, as well as some form of structured evaluation of IBST/L features at some point in the process. However, this model of PDP in which the teachers work in the same manner that they are about to instruct their pupils to work may be implemented even outside Finland.

When they attend a professional development programme, teachers are looking for something new. In a short-term programme, it is important to accept that each teacher is in her or his own phase of the development process. It would be very difficult to design teaching sequences that are suitable for all teachers—this would be likely to prove too revolutionary or too traditional. The most fruitful approach, then, may be to trust teachers and allow them to design their own IBST/L pilots, within a loose framework. During the PDP workshops, the teachers were introduced resources that would enable them to stimulate their own innovation process.

According to Sawyer (2004), 'Implementing creative teaching will require serious, long-term investment in professional development for teachers... Yet it has the potential to result in brighter, more motivated, and more effective teachers, and to result in students with deeper understanding and improved creative and social skills' (p. 18). It can be argued that the PDP succeeded in awakening participants to the realisation that there are alternative ways of organising teaching and that these may not be much more difficult to organise than traditional methods. The expectation was that, by empowering teachers to design inquiry pilots, their sense of responsibility for their own professional development would be bolstered and that the effects of this short-term PDP would become long term by introducing them to some novel approaches. In general, the teachers should be encouraged to adopt novel ways to do what they are doing anyway, not anything extra besides what they are already doing.

The aim here was to evaluate the potential of the professional development model by analysing teachers' presentations of their IBST/L pilots. This research did not explore students' conceptual learning or motivation. However, the potential to promote students' motivation is central to IBST/L instruction, which is why the topic was introduced to participating teachers and why it is included in the definition of inquiry. Every participant teacher succeeded in designing a pilot with at least some inquiry features. They also valued the collaborative discussions relating to the principles of IBST/L and the meaning of motivation in the classroom, as well as the reflective discussions of each other's IBST/L pilots.

References

Andersson, R. D. (2007). Inquiry as an organizing theme for science curricula. In S. K. Abell & N. G. Lederman (Eds.), *Handbook of research on science education* (pp. 807–830). London: Lawrence Erlbaum Associates.

Bybee, R. W. (2000). Teaching science as inquiry. In J. Minstrell & E. H. van Zee (Eds.), *Inquiring into inquiry learning and teaching in science* (pp. 20–46). Washington, DC: American Association for the Advancement of Science.

Donovan, M. S., Bransford, J. D., & Pellegrino, J. W. (1999). *How people learn: Bridging research and practice.* National Academy Press, 2101 Constitution Avenue NW, Lockbox 285, Washington DC, 2005.

Eccles, J. (2005). Subjective task-value and the Eccles et al. model of achievement-related choices. In A. J. Elliot & C. S. Dweck (Eds.), *Handbook of competence and motivation.* New York: Guilford Press.

Finnish National Board of Education (FNBE). (2014). *Perusopetuksen opetussuunnitelman perusteluonnos. A draft of the national core curriculum for basic education.* Helsinki, Finland: National Board of Education. Retrieved from http://www.oph.fi/ops2016 [in Finnish]

Guay, F., Ratelle, C., & Chanal, J. (2008). Optimal learning in optimal contexts: The role of self-determination in education. *Canadian Psychology, 49*(3), 233–240.

Jurow A. S., & McFadden, L. (2011). Disciplined Improvisation to extend young children's scientific thinking. In Sawyer, R. K. (Ed.), *Structure and improvisation in creative teaching* (pp. 236–251). Cambridge, NY: Cambridge University Press.

Juuti, K., Lavonen, J., Aksela, M., & Meisalo, V. (2009). Adoption of ICT in science education: A case study of communication channels in a teachers' professional development project. *Eurasia Journal of Mathematics, Science & Technology Education, 5*(2), 103–118.

Kim, M., Lavonen, J., Juuti, K., Holbrook, J., & Rannikmäe, M. (2013). Teacher's reflection of inquiry teaching in Finland before and during an in-service program: Examination by a progress model of collaborative reflection. *International Journal of Science and Mathematics Education, 11*(2), 359–383.

Krzywacki, H., Lavonen, J., & Juuti, K. (2015). There are no effective teachers in Finland—Only effective systems and professional teachers. In O.-S. Tan & W.-C. Liu (Eds.), *Teacher effectiveness: Capacity building in a complex learning era* (pp. 79–103). Andoven, MN: Cengage Learning.

Laursen, S., Liston, C., Thiry, H., & Graf, J. (2007). What good is a scientist in the classroom? Participant outcomes and program design features for a short-duration science outreach intervention in k-12 classrooms. *Life Sciences Education, 6,* 49–64.

Lavonen, J., Juuti, K., Aksela, M., & Meisalo, V. (2006). A professional development project for improving the use of ICT in science teaching. *Technology, Pedagogy and Education, 15*(2), 159–174.

Lumpe, A. (2007). Research-based professional development: Teachers engaged in professional learning communities. *Journal of Science Teacher Education, 18,* 125–128.

Mansvelder-Longayroux, D. D., Beijaard, D., & Verloop, N. (2007). The portfolio as a tool for stimulating reflection by student teachers. *Teaching and Teacher Education, 23*(1), 47–62.

Minner, D., Levy, A., & Century, J. (2010). Inquiry-based science instruction—What is it and does it matter? *Journal of Research in Science Teaching, 47*, 474–496.

Nelson, T. H. (2009). Teachers' collaborative inquiry and professional growth: Should we be optimistic? *Science Education, 93*(3), 548–580.

Niemiec, C. P., & Ryan, R. M. (2009). Autonomy, competence, and relatedness in the classroom. *Theory and Research in Education, 7*(2), 133–144.

Norris, N., Asplund, R., MacDonald, B., Schostak, J., & Zamorski, B. (1996). *An independent evaluation of comprehensive curriculum reform in Finland*. Helsinki, Finland: National Board of Education.

NRC. (2012). *A framework for K-12 science education: Practices, crosscutting concepts, and core ideas*. Washington, DC: The National Academies Press.

Patton, M. Q. (2002). *Qualitative research & evaluation methods* (3rd ed.). Thousand Oaks, CA: Sage.

Penuel, W. R., Fishman, B. J., Yamaguchi, R., & Gallagher, L. (2007). What makes professional development effective? Strategies that foster curriculum implementation. *American Educational Research Journal, 44*(4), 921–958.

Reeve, J., & Halusic, M. (2009). How K-12 teachers can put self-determination theory principles into practice. *Theory and Research in Education, 7*(2), 145–154.

Rodgers, C. (2002). Defining reflection: Another look at John Dewey and reflective thinking. *Teachers College Record, 104*(4), 842–856.

Ryan, R. M., & Deci, E. L. (2002). An overview of self-determination theory: An organismic-dialectical perspective. In E. L. Deci & R. M. Ryan (Eds.), *Handbook of self-determination research* (pp. 3–33). Rochester, NY: University of Rochester Press.

Sawyer, K. (2004). Creative teaching: Collaborative discussion as disciplined improvisation. *Educational Researcher, 33*(2), 12–20.

Sawyer, K. (2006). Educating for innovation. *Thinking Skills and Creativity, 1*, 41–48. https://doi.org/10.1016/j.tsc.2005.08.001

Taajamo, M., Puhakka, E., & Välijärvi, J. (2014). Opetuksen ja oppimisen kansainvälinen tutkimus TALIS 2013: Yläkoulun ensituloksia. Opetus- ja kulttuuriministeriön julkaisuja.

Tuominen-Soini, H. (2012). *Student motivation and well-being: Achievement goal orientation profiles, temporal stability, and academic and socio-emotional outcomes*. Doctoral dissertation University of Helsinki. Retrieved from http://urn.fi/URN:ISBN:978-952-10-8201-6

Uhrich, T. A. (2009). The hierarchy of reflective practice in physical education. *Reflective Practice, 10*(4), 501–512.

Vasalampi, K., Salmela-Aro, K., & Nurmi, J.-E. (2009). Adolescents' self-concordance, school engagement, and burnout predict their educational trajectories. *European Psychologist, 14*(4), 332–341.

Yeager, S. D. & Walton, G. M. (2011). Social-psychological interventions in education: They're not magic. *Review of Educational Research, 81*, 267–301.

Part IV
Capitalizing on Teacher Reflections to Enhance Inquiry-Based Science Teaching

The Biology Olympiad as a Resource and Inspiration for Inquiry-Based Science Teaching

Jan Petr, Miroslav Papáček, and Iva Stuchlíková

1 Introduction

Inquiry-based science teaching (IBST) is seen as important and is emphasized in current science education and education policy discourses, and yet there is some resistance towards teaching through inquiry (Wilcox, Kruse & Clough, 2015, p. 62). It could partly be the result of poor understanding of what IBST is and how it could be realized in the classroom (Demir & Abell, 2010). When preparing a teacher development programme (TPD) supporting inquiry-based science teaching, it is therefore necessary to address the possible misunderstandings, reservations, and hesitations of teachers. The preparation of a TPD programme must also reflect teachers' needs in terms of opportunities for theoretical understanding and provide a clear image of effective teaching and learning and also enough information about student learning (Birman, Desimone, Porter, & Garet, 2000; Borko, 2004; Guskey, 2002; Guskey & Yoon, 2009).

The main challenge during the development of an effective TPD programme aiming at IBST lies in the fact that the majority of teachers who do not teach through inquiry have only limited personal experience of this approach to teaching. According to a sociocultural perspective on teacher learning (Cobb, 1994; Lave & Wenger, 1998), it is important to provide teachers with experience, which could be seen as good practice in a newly introduced approach (Anderson, 2002; Sunal et al., 2001; Windschitl, 2003). A TPD programme for inquiry-based science teaching (IBST) should build upon teachers' previous participation in the learning and teaching culture in and out of school (Abell, Smith, & Volkmann 2004). The extracurricular activities called Science Olympiads and tasks created for these competitions

J. Petr (✉) · M. Papáček · I. Stuchlíková
Faculty of Education of the University of South Bohemia in České Budějovice,
České Budějovice, Czech Republic
e-mail: janpetr@pf.jcu.cz; papacek@pf.jcu.cz; stuchl@pf.jcu.cz

© Springer International Publishing AG, part of Springer Nature 2018
O. E. Tsivitanidou et al. (eds.), *Professional Development for Inquiry-Based Science Teaching and Learning*, Contributions from Science Education Research 5, https://doi.org/10.1007/978-3-319-91406-0_11

could be a promising example of good practice for inquiry-based science teaching, and teachers' experience of them could be an important reference for introducing and/or upscaling inquiry-based teaching.

Inquiry-based science teaching and education (IBSTE) is seen as an opportunity to improve students' understanding of key scientific concepts, the power of scientific methods, and interest in science-related careers (Rocard et al., 2007). These opportunities are understood and valued by teachers; but the teachers, of course, also see the limits, challenges, and obstacles. They see some challenges to the successful implementation of inquiry-based learning on the students' side, such as a lack of motivation, skills, and background knowledge. And, last but not least, they also consider practical constraints limiting inquiry teaching, e.g. resources, time, school timetable, etc. (Stuchlíková, 2010).

The concept of IBSTE has not been widely adopted in the Czech Republic (Stuchlíková, 2010) until recently, though its defining features are not brand new in the Czech education system. Some important facets of Inquiry-Based Science Education (IBSE) were accepted by Czech teachers earlier in different conceptual frames and influenced the Czech pedagogical practice and teacher training tradition. Thus other notions are seen as (at least) partly complementary to IBSE, e.g. problem-based learning, inductive thinking, critical thinking, experiential learning, and scientific methods of learning. Czech teachers are used to the concepts of problem-based teaching and project-based instructions and are trying to implement these approaches. On the contrary, concepts which became popular recently, such as heuristic methods, and especially IBST, seem rather distant to the teachers. It is therefore reasonable to bridge the gap between newly introduced concepts and teachers' current experience through something with which the teachers are currently familiar, albeit as an extracurricular activity. This could be, at least in the Czech context, the Science Olympiad (SciO). Anyway, to understand more deeply whether, and which, inspirations from SciO could be utilized in teacher education programmes systematic study is needed.

In this chapter, we summarize our investigation of the potential of the Biology Olympiad (BiO) to serve as a resource for inquiry-based teaching. We started with a survey of pre-service teachers' knowledge, beliefs, and experience to identify the main problems that pre-service teachers see in inquiry-based teaching and what sources of inspiration and good practice (including BiO) they have.

The next part of the chapter presents a study focused on the experience of BiO tasks and culture of those teachers who organized the competition in previous years. Using questionnaires, we collected data on their attitudes towards inquiry and on the impact that BiO has on their everyday instructional practice. To add to these findings, we also summarized the main findings from group discussions with teachers who participated in our summer schools on IBSE; we asked them what would help them to decide 'what and when' from BiO could be implemented in their everyday biology classes to inspire and motivate students towards inquiry.

As the teachers evaluate new practice from the perspective of whether it would improve the learning of their students (Guskey, 2002), we also collected data about secondary school students' views on Science Olympiads (in general). With the intention of providing participants in the TPD programme with detailed insights, we again combined the use of a questionnaire with in-depth interviews with students who have successfully participated over a longer period (beyond their school-level competition).

The teachers in our first studies generally appreciated the positive assets of BiO, but they saw the implementation of the BiO 'inspiration' in their everyday teaching practice as rather difficult and demanding; the final part of the chapter focuses on analysis of the possibilities of using the competition tasks in regular instruction.

Before presenting the research findings, we first introduce the Biology Olympiad itself and address the seeming contradiction of its competitive nature with the cooperative nature of IBSE.

2 Biology Olympiad as a Model for Inquiry-Based Science Education

Science Olympiads are a type of self-improvement competition, in which students carefully solve prepared complex tasks, which demand well-integrated knowledge, creativity, and scientific practices. These are organized in the subjects of chemistry, physics, and biology. The Biology Olympiad is a nationwide competition, which is organized as an extracurricular activity, albeit still under the auspices of the Czech Ministry of Education. The students are encouraged by their teachers to participate at the school level and possibly to continue further. BiO is a systematic and continuously organized activity, the aims of which are twofold: to motivate students and arouse their interest in science and to offer leisure-time activities for all students interested in biology and environmental issues. The competition has a funnel architecture, which means that the content is based on the school curriculum and the school level is accessible to almost all students interested in participating. On the other hand, the concept of BiO is congruent with the International Biology Olympiad (IBO), so that those students who advance to more difficult levels can take advantage of specialized camps and can also continue at the international level.

Every year, there are chosen competition topics, for which some preparatory texts are provided, so that the students (with the support of their teachers) can study and prepare on their own. Then, during the contest, the students are given complex tasks, both theoretical and practical (laboratory or field tasks). The students' solutions are thoroughly evaluated with respect to creative thinking, the use of evidence and scientific knowledge, the quality of their hypotheses and explanation, the correctness and elaboration of the conclusion, and the innovativeness of the solution. To standardize the process of evaluation, the group of scholars preparing the tasks also prepares criteria for evaluation and scoring.

BiO is in fact guided inquiry, where the learner must 'sharpen or clarify the question provided by the material' (National Research Council, 2000). This is followed by highly self-directed activity on the part of the student – he/she has to determine what constitutes evidence and collect it, to formulate an explanation based on evidence, to examine other resources and relate them to the explanation, and to formulate reasonable and logical arguments to communicate the authored solution. Thus, BiO meets the definition of inquiry-based science education, as it engages students in authentic, open-ended problem-based learning activities, in experimental procedures, experiments, and 'hands-on' activities, including the search for information, in self-regulated sequences of the application of knowledge and skills, and in argumentation and communication of the solution.

BiO has a positive influence on motivation and students' achievements and future professional orientation as well. Stazinski (1988) gives data from grammar schools which shows that 87% of competitors are still highly interested in biology 2 or 3 years after the competition. They are interested in biological literature, are very active in biology lessons, and achieve good results in biology. Their interest in biology is highly stable. In general, competitions provide an opportunity for the first or an early success that can help to attract the student to a scientific career (Kenderov, 2006).

On the evidence of feedback from participants in the European Science Olympiad (EUSO, a team competition for EU second-level school science students who are 16 years of age or younger), science competitions are accepted by students as a very enjoyable form of interaction with peers; students learned a lot about experimental science, gained new experiences, and felt that the competition had created a more positive view of science (O'Kennedy et al., 2005).

3   Competition in Science Education and Inquiry

Whilst communication with peers and working cooperatively in teams is frequently stated as an important feature of IBSTE (Flick & Lederman, 2004), it is necessary to explain the competitive nature of Science Olympiads and explore whether the competition itself can contribute to an inquiry-oriented classroom culture. At first sight, the idea of a competition seems to contradict the notion of inquiry. Competition has been contrasted in the educational context with cooperation and has usually been described as basically negative in recent decades. Anyway, the 'beauty and the beast' paradigm of cooperation and competition is nonetheless no longer tenable, and we have to use the advantages of both in science classes (Fülöp, 2009).

Even if we are aware of the negative sides of competition (a sense of threat, poor communication, suspicious and hostile attitudes, anxiety, fear of failure, interference with the cognitive functioning needed to solve problems), it is important to take into account research showing that constructive competition contributes to task solution effectiveness, personal benefits (such as social support), strong positive relationships, enjoyment of experience, the desire to participate, confidence in working collaboratively with competitors in the future, etc. (Tjosvold, Johnson, Johnson, & Sun, 2003). Sheridan and Williams (2006) described how constructive competitive relationships are constituted and relate to (even preschool children's) motivation and learning. Olympiad-like forms of competition also seem to contribute in many ways to an inquiry-oriented classroom culture as they are constructive, helping the students to share and discuss science-related ideas.

It is important to say that competition can be a friendly process in which the competing parties mutually motivate and improve each other, but it can also be a desperate fight full of aggression and frustration. However, the idea of the Olympiad is first and foremost about fair stakes ('it is not important to win but to participate'). Adolescent students naturally also value such kinds of competition as they need increasingly precise feedback concerning their own school performance, as well as help in building their own self-esteem. To participate in an Olympiad is an activity which is valued by peers. Olympiads offer the option to participate and thus receive additional support from other students who participate (teammates) and from teachers and peers, which may in turn serve to upgrade their social status in the classroom.

The fair competitive context also somewhat simplifies the learning context as the tasks provide a structure which can enable students to temporarily neglect other non-task-related classroom goals (e.g. social preferences, etc.). Competition also explicitly opens the door for performance-oriented goals. In pedagogical practice, it is especially difficult to separate learning goals that are intrinsically motivated from extrinsically motivated performance goals, as they are often combined with individuals' motivation. The combinations of learning goals may determine how the students approach their tasks. Competition satisfies striving for multiple learning goals – it combines an emphasis on both mastery and performance, which is most beneficial for students (Barron & Harackiewicz, 2001; Schunk, Pintrich, & Meece, 2008). Another aspect valued by students is that the Olympiad competition provides all participants with detailed feedback on their performance and current state of knowledge and skills[1] and enables comparison with other students at similar levels of achievement, so they can view their own performance from a broader perspective. This applies to all participants, from the school level up to the local, regional, or national level.

Even for those with lower abilities or lower self-esteem, or for those who refuse to participate, there may be some benefits emanating from social comparison. Social comparison is one of the major features of the classroom environment. Adolescents have a tendency to nominate a comparison target from the same-sex students who slightly outperform them in the class, who serves as a means of self-improvement (Huguet, Dumas, Monteil, & Genestoux, 2001). Additionally, witnessing the successful striving of other people may provide vicarious experience of success, which is even more important when the person who acts as a model is in a close relationship with the observer (Bandura, 1997). Hence, the shared positive experience may encourage even those who are not immediately participating. It is, of course, an indirect influence, but it is nonetheless reasonable to draw on these 'side effects' of Olympiads.

---

[1] The participants receive a detailed assessment of their solution; feedback is provided by the organizing teachers or scholars who prepare the task and formulate the assessment criteria.

4   Pre-service Teacher's Views on Inquiry and Biology Olympiad

Bearing in mind that pedagogical constructivism is the theoretical basis underlying the development of students' critical and inquiring thinking, we explored future teachers' knowledge of constructivist ideas.

We addressed 160 (142 females, 18 male) pre-service biology teachers and students of 4 faculties from 2 Czech universities located in different regions. The open-ended questionnaire that was constructed ad hoc asked about the content and connotations of the term 'constructivism', and the respondents' potential ability to use the heuristic method of teaching was assessed by means of two tasks, in which the respondents ought to design laboratory experiments for two biological phenomena – photosynthesis and digestion. Kolčavová (2011) surveyed the main results of this study as follows: only less than half of the respondents (42%) were familiar with the term 'constructivism' and were able to explain correctly its core ideas and characterize the main features of this pedagogical approach. Contrary to these results, nearly all the respondents were able to prepare some school experiments and use them knowingly or intuitively for the heuristic teaching of the aforementioned biological phenomena. The results of this investigation showed some indication that the women were slightly better at dealing with pedagogical terminology (44% of the women, as compared with 28% of the men, were able to explain the terms) and, conversely, the men were slightly better at inventing school experiments (90% of the men, as compared with 75% of the women, were able to design experimental procedures to illustrate or explore these biological phenomena). Because of the small number of men involved, however, the difference was not statistically significant.

In addition to this part of the research, we asked 24 respondents from the original sample for an interview about their motivation and sources of inspiration for the preparation and use of inquiry tasks in future teaching. Their most frequent answers can be roughly summarized as follows: '… because I enjoy it; because it satisfies what I think a teacher should do', but also '… because I want to be good in the classroom and I want to motivate my pupils; or … for a better understanding of my pupils'. As sources of inspiration, they referred to experiences and examples from their own education, textbooks, or the Internet but also referred relatively frequently to BiO tasks.

This part of the investigation can be summarized by stating that pre-service teachers' potential to change their pedagogical thinking and apply IBST in their future teaching practice is probably better than one would infer from their level of knowledge of constructivist teaching (see above). Nevertheless, it is necessary to develop the knowledge, skills, and attitudes that they will need for enactments of inquiry teaching. But first of all it means providing them with experience of teaching through inquiry in practice (see also Guskey, 2002; Klein, 2004). BiO could be seen as an example of a good practice option.

5   In-Service Teachers' Experience with Biology Olympiad and Their View on IBST

The second set of phenomena to be investigated was therefore the attitudes of teachers who are involved in the organization of BiO and in coaching participating students (N = 74). We wanted to know (via questionnaire and interviews) whether this experience of BiO can have a formative influence on willingness to teach through inquiry. The respondents were both pre- and in-service teachers of biology from two regions of the Czech Republic. The in-service teachers had varying lengths of teaching experience (from 1 to 25 years) and worked at secondary schools and gymnasiums from cities of different sizes (from about 5 thousand to more than 1 million citizens). An important feature of the teachers who participated in organizing BiO is that biology teaching was their main interest and some aspect of biology was a hobby for the majority of them.

The main findings can be summarized as follows. On the whole, these teachers do not consider BiO a regular part of biology education in school. They perceive it as an extraordinary and voluntary part of their professional activities. Nevertheless, organizing participation in the Olympiads, and coaching students, is sometimes supported by the school management. Only a few of the teachers who were interviewed, who are BiO organizers and/or coaches, work on their own. Usually, the teachers cooperate and work in school or regional teams. Some teachers pointed out the beneficial effect that the competition tasks have upon their teaching, mostly in the area of motivating students and the preparation of tasks for students' experiments, practice, and projects. But the teachers also stated that the transformation of BiO tasks into regular classroom activities for the biology curriculum is very difficult, and they named numerous barriers in this respect. Secondary school teachers especially consider simplified BiO tasks too time-consuming and demanding. They are also afraid of possible constraints, as most of their pupils are not particularly interested in or motivated for biology and so the tasks may not appeal to them. These problems are often mentioned by other authors (e.g. Eastwell, 2009). The barriers, and possible ways to overcome them, must be thoroughly addressed in teacher training (Foss & Kleinsasser, 1996; Guskey, 2002; Klein, 2004) in order to change teachers' attitudes and increase their willingness to use inquiry-based teaching.

Most teachers are not familiar with the notion of inquiry-based science teaching (they were and still are not systematically trained to implement this method of teaching). But they are able to use the inspiration from the Biology Olympiad (motivation, task culture, etc.) for the more or less intuitive support of inquiry-oriented facets of their teaching.

On the other hand, most pre- and in-service teachers (70% of all respondents) would embrace a course dealing with IBSE if it were available in pre- and postgraduate professional development programmes for teachers. A slightly smaller number of respondents (ca. 66%) pointed out that they would appreciate a course dealing with BiO tasks (how to create them and elaborate the assessment) and their implementation for fostering inquiry-based biology education. These findings seem promising. Nevertheless, the sample of respondents that was addressed was special as it consisted of teachers who are influenced by the BiO 'culture' and are somewhat more motivated and oriented towards heuristic teaching and inquiry-based biology than biology teachers on average.

Furthermore, we also organized two summer schools of inquiry-based biology education (N = 64 participants) for pre- and in-service teachers' professional development and for teacher educators. During the final discussions, we also organized group discussions on various aspects of IBSE related to BiO tasks. Out of these, two most issues were most salient:

1. How to decide which BiO tasks to choose, which are usable in classrooms, and which will increase students' motivation? The conclusion of the group discussion was to choose the tasks on the basis of classroom experiments and positive experiences of them. The tasks could be applied through in-class competition between groups, but the competition is not necessary, as the tasks are stimulating per se. Their main contribution lies in the development of students' science practices as the tasks have attractive content and are cognitively activating;
2. A desire for ready-made 'operating instructions' for tasks that teachers could automatically integrate (copy) into their teaching. Though understandable, this requirement is not reasonable, but it reflects the still superficial understanding of IBST of the participating teachers.

As a summary of this part of the investigation, we can conclude that BiO tasks, when suitable for classroom experiments, could be used for implementing inquiry-based biology teaching in regular classroom instructions. Teacher training for inquiry-based biology teaching nonetheless has to go 'beneath' the tasks and has to be based on deepened content knowledge and skills to respond properly to the changing environment in the classroom (see, e.g. Abell et al., 2004; Schwarz & Crawford, 2004).

6 Secondary Students' View on Olympiads

We interviewed 104 secondary school (ISCED 2 and 3) students (two students in grades 9–12 per school from the South Bohemia region) about their prospective participation in the Olympiads and about the incentives from the teacher, school or school community, and family that could foster their willingness to take an active part.

Thirty-eight students (37%) had not participated in any Olympiad or similar competition. These students described the reasons as perceived lack of knowledge, skills, or motivation. Some of them dislike competitive situations; some would prefer teamwork to individual competition. The fear of failure in comparison with others and anticipated poor results blocked some of these students. When asked about prospective support, the students wanted more assistance from the teacher; some needed to be assured that they would be able to cope with the tasks; some would like to get support from classmates and would like to have some continuous, all-year-round activities in order to prepare.

Sixty-six students (63%) had participated in Olympiads (one or more). Their decision to participate was a consequence of previous interest, of pleasure in competitive activities, and of a willingness to learn more about science or prepare for a career in science. But the most frequently mentioned reason was the teacher's stimulation or encouragement. Generally, the most effective stimulation is open and sincere assessment of the student's potential to master more and to solve puzzling and demanding tasks. This corresponds with research findings on teachers' roles in classroom goal orientation (Roeser, Marachi, & Gehlbach, 2002; Bong, 2001).

What the students especially appreciated, and would like to see introduced into classroom lessons, were the interesting competition tasks, which were more related to everyday life and more based on hands-on activities and provided enough time to 'build up' the solution (allowing time for errors when searching for solutions).

7   Transforming the Ideas of the Biological Olympiad into Classroom Practice

The Biology Olympiad is a competition with a long tradition. On the national level, it has already been held for 50 years, and about 60 countries are now involved. The competition has hard-and-fast rules. It is held at three national levels and one international level, and students and pupils compete in four age groups (Farkač & Božková, 2006; MEYS, 2007).

Tasks are developed for the school, regional, and national levels. The portfolio of tasks produced for BiO is relatively broad. Some tasks are very similar to regular school tasks (primarily at the school competition level). The use of these tasks in classroom lessons is possible, but they provide only minor additional IBSE benefits (e.g. simple observation, multiple choice tests, crosswords, picture description, etc.). This kind of task is used for further training in some skills and types of knowledge at the school level.

Newly designed competition tasks are the main source of enrichment for biology instruction. The tasks produced over the last few years are more inquiry-oriented, or they have at least some inquiry-related features. Tasks of that kind can change transmissive instructional methods and extend teachers' portfolios of instructional methods and forms (e.g. complex tasks based on experiments where several inquiry skills are needed; see in Appendix).

BiO is a potential source of enlivenment of biology instruction and fostering inquiry-based teaching via the application of competition tasks. Teachers can get a set of tasks which are produced by teams of specialists from different branches of biology. Thanks to the participation of teachers and biology students when the tasks are prepared, both their factual and didactic qualities are guaranteed. The detailing of the tasks continues for as long as possible before the competition, so the final manual and instructions for teachers and the jury are published last of all. Teachers can obtain an excellent source of new and well-elaborated tasks that can inspire their practice. In addition to many simple tasks for particular science practice training, there are approximately 15 complex theoretical and laboratory tasks developed for 1 year of the competition. Therefore, many new tasks are created which could potentially be used in instruction. Teachers obtain well-elaborated methodological instructions and elaborated authors' solutions for every task showing what potential for IBST the task has. The solution of a BiO task is a tool for the evaluation of the tasks, on the one hand, and for the evaluation of educational processes on the other. Evaluation of the tasks from the point of view of their success rate is also a good indicator for the potential implementation of the selected task in a particular class context.

There are also some limits on the implementation of the competition tasks into classroom instruction. BiO mainly aims at interested or gifted students and therefore has an extracurricular character (i.e. the task may have a different thematic scope in comparison with the standard curriculum). For this reason, some tasks require extracurricular knowledge. Special equipment or material could also be required, and it is usually provided only for the relevant level of the competition by the central competition committee.

Some tasks, primarily from higher levels of the competition, must be adapted by teachers for use with less advanced students, because they are too difficult or time-consuming. Most tasks require a greater or lesser adjustment to inquiry-based teaching in everyday lessons, e.g. matching the relevant curriculum, specification of a hypothesis, format of the presentation of results, etc.

7.1 Coherence/Links Between Main Features of IBSTE and BIO Tasks

The set of competition tasks used in BiO in the Czech Republic during the previous 15 years was studied. A model sample of these tasks was published with commentaries by Petr (2014). It is mostly possible to classify the tasks as structured or directed inquiry (according to the classification of Eastwell, 2009). Students at the lower secondary level are registered into the competition with their paper (called an 'entry task'), which can be based on open or directed inquiry. Altogether, about 240 competition tasks were analysed from the point of view of individual levels of inquiry (according to Eastwell, 2009). The main evaluation criterion was the occurrence of basic and integrated process skills necessary for the development of scientific thinking as defined by the American Association for the Advancement of Science (AAAS, 1989). The results of the analysis show significant overlap and coherence of their essential attributes with scientific processing skills (Table 1). It is

Table 1 Attributes of BiO tasks in relation to key characteristics of IBSTE

Levels of inquiry according Eastwell (2009)	BiO tasks	Stages of inquiry cycle (Justice et al., 2007, modified)	BiO tasks	Skills used in inquiry (according AAAS, 1989)	BiO tasks
Level 1 (confirmation)	Yes	Questions	Yes	Observation	Yes
Level 2 (structured)	Yes	Hypotheses	No	Measurement	Yes
Level 3 (directed)	Yes	Designing of simple experiments	Yes	Classification	Yes
Level 4 (open)	Yes[a]	Data collection	Yes	Inferring	Yes
		Understanding	Yes	Predicting	Yes
		Discussion, communication	No	Identifying variables, relationships	Yes
				Communication	No
				Interpreting data	Yes
				Controlling variables	No
				Operational definitions	No
				Hypothesizing	No
				Experimenting	Yes

[a]Only at the lower secondary level, entry tasks can be based on open inquiry

evident that the BiO tasks are suitable for the development of inquiry skills and science competencies.

7.2 What Features of the BiO Tasks Are Especially Suitable for IBSE?

The competition tasks proved their potential to improve inquiry science education. The Olympiad tasks resemble real research more closely (Kenderov, 2006). For example, Breyfogle (2003) stated that within the Chemistry Olympiad the emphasis is placed on problem-solving skills and hands-on and minds-on constructivist learning practices, which is not typical of the daily school chemistry laboratory setting.

It is similar in biology teaching, with one objection. Whilst chemical experiments are mostly very attractive and fast and show an immediate effect, biological experiments are slower and the effect is often less evident. Therefore, newly designed competition tasks based on simple experiments are very valuable. BiO produced just these tasks.

Within BiO, there are two main kinds of competition tasks suitable for transfer into inquiry classes and lab work in school:

1. Theoretical tasks without the requirement for any special laboratory equipment. Nonetheless, these tasks do not lead only to the identification of bare facts; they are complex and sophisticated and are not solvable without complex

problem-solving operations and proper work with data. Additionally, the verification of the inferred solution is realized in different ways (filling in missing information, content analysis of texts, work with pictures, tables, diagrams, etc.).

2. Laboratory tasks which are designed to use basic laboratory or field equipment. In addition to tasks based on simple observation, laboratory procedures, and measurement, more complex experimental laboratory tasks are used within BiO. Generally, they are not very complicated, but it is necessary to redesign some of them. Complicated tasks or tasks from higher levels have to be simplified for younger pupils. On the other hand, it is possible to refine or to extend relatively simple tasks, depending on the curricular content.

Examples of these two types of tasks from BiO are presented with evaluations and examples of their implementation in regular classroom instruction in the Appendix.

7.3 Possibilities of Incorporating BiO into Teacher Education

Several possible ways to use the Science Olympiad in teacher education were mentioned by Breyfogle (2003). On the basis of his experience, pre-service teachers can contribute to the design of competitive events. For example, they can prepare rule sheets and design guided inquiry laboratory activities, questions, etc. Pre-service teachers mentioned very good experience of preparing and writing lab activities, supervising the competition, and working with students.

We positively tested the following possibilities for incorporating BiO into teachers' education:

1. The first possibility is the participation of pre-service teachers in the school or regional levels of the competition. Pre-service teachers can work on the jury, can assist with laboratory hardware, and can prepare materials, but they can especially observe the activity of the competitors, the level of their knowledge, their way of solving the tasks, etc. Our pre-service student teachers saw this participation as a great benefit for their prospective inquiry-based biology teaching.

2. The tasks from BiO are first-rate material for content and didactic task analysis which can be directed towards their selected characteristics – level of difficulty, the development of educational competencies, demands on intellectual operations, and their potential for deductive or inquiry-based science education. The most effective way to offer BiO tasks to teachers for regular instruction is through seminars, where, after working on the selected task in groups, the participants thoroughly discuss and evaluate not only the tasks but also related epistemological beliefs, teaching and management strategies, questions related to the benefits students derive from the tasks (e.g. autonomy, working in heuristic circles, cooperation vs. competition, etc.), possible limitations on their application in their particular school, etc. Almost all the pre-service and in-service teachers who participated in the seminars supported by S-TEAM considered this way of

working with tasks to be very useful and helpful, as well as a good resource for introducing inquiry-based methods into their own practice.

8   Conclusion

The effectiveness of a professional development programme for teachers to support new practice depends heavily on the correct assessment of the teachers' current needs and of their preparedness to progressively master new practice (Guskey & Yoon, 2009). The first phase of the preparation of an evidence-based TPD programme therefore comprises research focused on an individual professional development programme at a single site. A possible component of the professional development programme must be considered with respect to teachers as learners (Borko, 2004). The investigation presented in this chapter was a part of such an endeavour.

The findings of our studies have shown that Czech teachers are not yet familiar with the concept of inquiry-based science teaching, but value its potential for their learning. As there are strong expectations concerning IBST on one side, as well as hesitation on the other, it is important to search for current teachers' experience which can represent a model of IBSE. We have found such a body of experience within the extracurricular activity involved in the Biology Olympiad. Key aspects of inquiry-based science teaching can be introduced through the BiO tasks. Positive attitudes of pre-service teachers and especially of those in-service teachers who participated in the organization of BiO towards the possible transfer of some BiO tasks into everyday classroom practice led us to the question of what type of tasks could be transferred and under which conditions.

The secondary students' view of Olympiads as something interesting and valued by their peers is promising for improving students' motivation via the implementation of BiO tasks and makes their adoption by teachers more plausible. The prospects for the success of transferring the inquiry nature of BiO and its tasks into everyday biology teaching is nonetheless dependent on further elaboration of the relevant teachers' pedagogical content knowledge, skills, and attitudes towards inquiry.

Appendix: Examples of Competition Tasks with Comments

Example 1

This task is focused on genetic drift, which is a powerful microevolutionary mechanism and moves with the frequency of alleles in populations (Hájek, 2005).

Task: genetic drift is a random process. To illustrate genetic drift better, we will simulate it via a lottery game. Your hypothetical population has ten subjects.

Everyone will produce two gametes. Therefore, 20 gametes altogether are produced in every generation. But only ten gametes come down to the next generation. For simplicity: we will expose a gene with two alleles to the drift effect. We will simulate these alleles with paper cards or some stones in two colours. We will start from a ratio of alleles (cards/stones) of 10:10 in the first part and 4:16 in the second part of the game. Every part will have five rounds. Because of the design of the game, the results can be quite different in comparison with your expectations. Let us allow ourselves to be surprised.

Part 1

You have a lottery device (envelope, bag, or the like) and cards/stones in two colours (a total of 20 pieces, 10 of each colour). Put 20 cards into the lottery device in every game round. In the first round, put ten cards of both colours into the device. Mix and toss ten cards. Write the number of cards, multiply the numbers of each colour by two, put into the device, and toss once more. Do this for five rounds and write the results into the table (write the colours of the cards as well).

Part 2

Repeat the game, but put 4 cards of one colour and 16 cards of the second colour into the lottery device at the beginning (the ratio of the alleles is different this time). Toss five rounds again and record the outcomes.

Questions

a. What ratio of alleles was there at the beginning of the game, i.e. before the simulation of genetic drift? Specify in %.
b. What ratio of alleles was there at the end of the game, i.e. after the simulation of genetic drift? Specify in %.
c. It is possible that you have managed to eliminate one of the two alleles wholly. The second setting of the game is more inclined to the elimination of one of the two alleles. Why?
d. The term 'fixation of alleles' is relevant to the elimination of an allele from the population. What does this term mean?

The advantages of this task: (1) it is very simple; (2) relatively difficult phenomenon is illustrated; (3) work in small groups or in pairs is possible (support for teamwork); (4) almost no equipment is necessary; (5) the possibilities for follow-up activities (statistics, etc.) are evident; and (6) there is a high motivational effect and a possibility of discussion.

The disadvantage of this task: it is a relatively time-consuming task; it is more suitable for older students (knowledge of genetics and their terminology is required).

Example 2

This task is focused on human physiology, specifically the physiology of digestion (Team of authors, 1997).

Task: There are some pieces of food in front of you (e.g., a roll, bread, an apple, cheese, a carrot). Taste small pieces of food one by one – put them into your mouth, chew thoroughly, and leave them for a short time in the oral cavity.

1. Observe which foods have changed their taste. Characterize this change.
2. By which process was the change of taste caused? Which enzyme is responsible for this change?
3. Observe also how the pH in your oral cavity will change during the consumption of food. Take the pH before eating and immediately after consuming all the pieces of food. Afterwards, students must pair up – one of you will chew some chewing gum without sugar and will observe if the chewing influences the pH level of the oral cavity as in TV commercials. The second individual, the control, will measure the pH at the same intervals without chewing gum. Log the pH values immediately after eating, after 5 min, after 10 min, and subsequently after every 30 min (a minimum of three times altogether). Write your outcomes into the table and draw it up in a graph.

Note: the necessary hygienic rules, methods for measuring pH, an empty table, and several additional tasks are an integral part of the instructions for this task.

Commentary

This task is a suitable example of the application of competition tasks in IBSE. It starts with an experimental component. On the basis of this very simple experiment, a student can continuously record data, and afterwards the data may be represented in the form of a graph. It is necessary to interpret the acquired data correctly in connection with the wider biological context.

The advantages of this task: (1) it is very simple; (2) it has an association with everyday life; (3) it is simple example of the physiology of digestive systems; (4) work in small groups or in pairs is required (support for teamwork); (5) although this task is designated for a higher category, it is possible to use it at lower educational levels without intensive adaptation; (6) only minimal laboratory equipment is necessary; (7) the possibilities for follow-up activities (statistics, etc.) are evident; and (8) there is a high motivational effect and the possibility of discussion about TV commercials.

The disadvantage of this task: it is a relatively time-consuming task, so it is more suitable for longer practices or homework (and follow-up work with data in the school).

References

Abell, S. K., Smith, D. C., & Volkmann, M. J. (2004). Inquiry in science teacher education. In L. B. Flick & N. G. Lederman (Eds.), *Science inquiry and nature of science. Implications for teaching, learning, and teacher education* (pp. 173–200). Dordrecht: Kluwer Academic Publisher.

American Association for the Advancement of Science. (1989). *Project 2061: Science for all Americans.* Washington, AAAS. http://www.project2061.org/publications/sfaa/online/sfaatoc.htm

Anderson, R. (2002). Reforming science teaching. What research says about inquiry? *Journal of Science Teacher Education, 13,* 1–12.

Bandura, A. (1997). *Self-efficacy: The exercise of control.* New York: Freeman.

Barron, K. E., & Harackiewicz, J. M. (2001). Achievement goals and optimal motivation: Testing multiple goal models. *Journal of Personality and Social Psychology, 80,* 706–722.

Birman, B. F., Desimone, L., Porter, A. C., & Garet, M. S. (2000). Designing professional development that works. *Educational Leadership, 57,* 28–33.

Bong, M. (2001). Between- and within-domain relations of academic motivation among middle and high school students: Self-efficacy, task-value, and achievement goals. *Journal of Educational Psychology, 93,* 23–34.

Borko, H. (2004). Professional development and teacher learning: Mapping the terrain. *Educational Researcher, 33*(8), 3–15.

Breyfogle, B. E. (2003). Using the science Olympiad to prepare preservice chemistry teachers. *Journal of Chemical Education, 80*(10), 1165–1167.

Cobb, P. (1994). Where is the mind? Constructivist and sociocultural perspectives on mathematical development. *Educational Researcher, 23*(7), 13–20.

Demir, A., & Abell, S. K. (2010). Views of inquiry: Mismatches between views of science education faculty and students of an alternative certification program. *Journal of Research in Science Teaching, 47*(6), 716–741.

Eastwell, P. (2009). Inquiry learning: Elements of confusion and frustration. *The American Biology Teacher, 71*(5), 263–264.

Farkač, J., & Božková, H. (2006). *Biologická olympiáda* [The biology Olympiad]. Prague: Jan Farkač.

Flick, L. B., & Lederman, N. G. (2004). *Scientific inquiry and nature of science implications for teaching, learning, and teacher education.* Dordrecht: Kluwer Academic Publishers.

Foss, D., & Kleinsasser, R. (1996). Pre-service elementary teachers' views of pedagogical and mathematical content knowledge. *Teaching and Teacher Education, 12*(4), 429–442.

Fülöp, M. (2009). Happy and unhappy competitors: What makes the difference? *Psychological Topics, 2,* 345–367.

Guskey, T. R. (2002). Professional development and teacher change. *Teachers and Teaching: Theory and Practice, 8*(3–4), 381–391.

Guskey, T. R., & Yoon, K. S. (2009). What works in professional development? *Phi Delta Kappan, 90,* 495–500.

Hájek, J. (2005). *Biologická olympiáda 2004–2005. 39. ročník. Zadání soutěžních úkolů. Krajské kolo kategorie A, B.* [The biology Olympiad 2004–2005. 39th Year. The Regional Level of Category A, B. Praha, The Czech Republic: Institute of Kids and Youth. Ministry of Education, Youth and Sport of the Czech Republic.

Huguet, P., Dumas, F., Monteil, J. M., & Genestoux, N. (2001). Social comparison choices in the classroom: Further evidence for students' upward comparison tendency and its beneficial impact on performance. *European Journal of Social Psychology, 31*(5), 557–578.

Justice, C., Rice, J., Warry, W., Inglis, S., Miller, S. & Sammon, S. (2007). Inquiry in higher education: Reflections and directions on course design and teaching methods. *Innovative Higher Education, 31*(4), 201–214.

Kenderov, P. S. (2006). *Competitions and mathematics education.* In Proceedings of the International Congress of Mathematicians (pp. 1583–1598). Madrid, Spain: European Mathematical Society

Klein, M. (2004). The premise and promise of inquiry based mathematics in pre-service teacher education: A poststructuralist analysis. *Asia-Pacific Journal of Teacher Education, 32*, 35–47.

Kolčavová, Z. (2011). *Připravenost studentů učitelství přírodopisu a biologie na interaktivní konstruktivistické vyučování s využitím prekonceptů žáků* [Biology and science pre-service teachers' readiness to interactive constructivist teaching using learner's preconceptions]. Unpublished master's thesis. České Budějovice, The Czech Republic: University of South Bohemia in České Budějovice.

Lave, J., & Wenger, E. (1998). *Communities of practice: Learning, meaning, and identity.* Cambridge: Cambridge University Press.

Ministry of Education, Youth and Sport of the Czech Republic. (2007). Organization system of the biological Olympiad. *Bulletin of Ministry of Education, Youth and Sport of the Czech Republic, 63*(7), 28–32.

National Research Council. (2000). *Inquiry and the national science education standards: A guide for teaching and learning.* Washington, D.C.: National Academies Press.

O'Kennedy, R., Burke, M., van Kampen, P., James, P., Cotter, M., Brown, W. R., et al. (2005). The first EU science Olympiad (EUSO): A model for science education. *Journal of Biological Education, 39*(2), 58–61.

Petr, J. (2014). *Možnosti využití úloh z Biologické olympiády ve výuce přírodopisu a biologie* [Possibilities of the Use of Tasks Produced within the Frame of the Biology Olympiad in the Natural Science and Biology Instruction]. České Budějovice, The Czech Republic: University of South Bohemia in České Budějovice.

Rocard, M., Csermely, P., Jorde, D., Lenzen, D., Walberg-Henriksson, H., & Hemmo, V. (2007). *Science education now: A renewed pedagogy for the future of Europe* (1st ed.). Luxembourg: Office for Official Publications of the European Communities. http://ec.europa.eu/research/science-society/document_library/pdf_06/report-rocard-on-science-education_en.pdf.

Roeser, R. W., Marachi, R., & Gehlbach, H. (2002). A goal theory perspective on teachers' professional identities and the contexts of teaching. In C. Midgley (Ed.), *Goals, goals structures, and patterns of adaptive learning* (pp. 205–241). Mahwah, NJ: Erlbaum.

Schunk, D. H., Pintrich, P. R., & Meece, J. L. (2008). *Motivation in education: Theory, research, and applications* (3rd ed.). Upper Saddle River, NJ: Pearson/Merrill Prentice Hall.

Schwarz, R. S., & Crawford, B. A. (2004). Authentic scientific inquiry as context for teaching nature of science: Identifying critical elements for success. In L. B. Flick & N. G. Lederman (Eds.), *Science inquiry and nature of science. Implications for teaching, learning, and teacher education* (pp. 331–356). Dordrecht: Kluwer Academic Publisher.

Sheridan, S., & Williams, P. (2006). Constructive competition in preschool. *Journal of Early Childhood Research, 4*(3), 291–310.

Stazinski, W. (1988). Biological competitions and biological Olympiads as a means of developing students interest in biology. *International Journal of Science Education, 10*(2), 171–177.

Stuchlíková, I. (2010). Concept of inquiry- based learning in teachers' professional development. In T. Janik & P. Knech (Eds.), *New pathways in professional development of teachers* (pp. 195–201). Vienna: LIT Verlag.

Sunal, D., Hodges, J., Sunal, C., Whitaker, K., Freeman, L., Edwards, L., et al. (2001). Teaching science in higher education: Faculty professional development and barriers to change. *School Science and Mathematics, 101*(5), 246–257.

Team of authors. (1997). *Biologická olympiáda 1996 – 1997. 31. ročník. Zadání soutěžních úkolů kategorie A.* [The biology Olympiad. Instructions and competition tasks. 31st year. Regional level of category A]. Prague: Institute of Children and Youth of the Ministry of Education, Youth and Sport of the Czech Republic.

Tjosvold, D., Johnson, D. W., Johnson, R. T., & Sun, H. (2003). Can interpersonal competition be constructive within organizations? *Journal of Psychology, 137*(1), 63–64.

Wilcox, J., Kruse, J. V., & Clough, M. P. (2015). Teaching science through inquiry: Seven common myths about this time-honored approach. *The Science Teacher, 82*(6), 62–67.

Windschitl, M. (2003). Inquiry projects in science teacher education: What can investigative experiences reveal about teacher thinking and eventual classroom practice? *Science Education, 87*, 112–143.

A Teacher Professional Development Programme on Dialogic Inquiry

Margareta Enghag, Susanne Engström, and Birgitta Norberg Brorsson

1 Introduction

In this chapter we present a pilot teacher professional development programme (TPDP) intended to give teachers opportunities to learn how to analyse classroom activities and to make changes needed in order to create a classroom characterised by dialogic inquiry-based science teaching. In the following section, we present dialogic inquiry in the classroom context.

1.1 Dialogic Inquiry as a Means to Enhance Science Teaching and Learning

The development of a dialogic science classroom is a way to enhance teaching and learning, where students of different backgrounds and different languages and experiences create the classroom environment (e.g. see Martin, 1993; Mortimer & Scott, 2003; Wells, 1999). There were several challenges that had to be taken into consideration in the development of the science teaching and learning in a Swedish school context. One challenge was the new Swedish curriculum and the new grading system (Swedish National School Agency, 2012) implemented during the S-TEAM project period but already advanced to take in new suggestions from

M. Enghag (✉)
Department of Mathematics and Science Education, Stockholm University,
Stockholm, Sweden
e-mail: margareta.enghag@mnd.su.se

S. Engström · B. N. Brorsson
School of Education, Culture and Communication, Mälardalen University, Vasteras, Sweden
e-mail: sengstro@kth.se; birgitta.brorsson@mdh.se

© Springer International Publishing AG, part of Springer Nature 2018
O. E. Tsivitanidou et al. (eds.), *Professional Development for Inquiry-Based Science Teaching and Learning*, Contributions from Science Education Research 5, https://doi.org/10.1007/978-3-319-91406-0_12

S-TEAM. Another challenge was the large proportion of multilingual students, who need to learn science and a new language simultaneously (Kouns, 2014; Swedish National School Agency, 2012). Yet another was the frequent lack of adequate subject knowledge among science teachers (Ottander & Ekborg, 2012) and a lack of professional development opportunities. So, our society of diversity called for new solutions and more focus on language and communication as a prerequisite for learning. Dialog can be seen as a meaning-making process of inter-thinking (Mercer, 1995) between teacher and students when specific questions are raised, which could be fruitful as steps towards inquiry. One way of developing a dialogic science classroom is therefore to use dialogic inquiry. Inquiry is defined in different ways. One definition we find useful is from Linn, Davis and Dell (2004, p. 4):

> We define inquiry as the intentional process of diagnosing problems, critiquing experiments, distinguishing alternatives, planning investigations, researching conjectures, searching for information, constructing models, debating with peers and forming coherent arguments often with technology resources.

Another definition of inquiry comes from Edelson, Gordin and Pea (1999, p. 393), who concluded in one of the first articles about IBST/L that the three learning objectives were (1) general inquiry abilities, (2) specific investigation skills and (3) understanding of science concepts and principles. We argue that these are interesting principals for dialogic inquiry too. Gyllenpalm, Wickström and Holmberg (2010) describe inquiry as (a) learning to do inquiry, (b) learning about inquiry and (c) learning science subject matter, which support Edelson et al. totally.

Dialogic inquiry does not refer to a method, but as Wells (1999, p. 121) expresses it in the book *Dialogic Inquiry*:

> …it indicates a stance towards experiences and ideas, a willingness to wonder, to ask questions, and to seek to understand by collaborating with others in the attempt to make answers to them.

From our perspective, inquiry is said to promote leaving the teacher-centred instruction for a student-centred instruction. Inquiry-based science teaching/learning (IBST/L) can be organised into several degrees of teacher involvement from guided inquiry to open inquiry (Guisti, 2008). This invites a more sociocultural view of teaching and learning, which will explain the difference between inquiry and *dialogic* inquiry. By *dialogic* inquiry we mean stressing the importance of dialog between peers, dialog between teacher and students and the use of writing as a tool in creating dialog (Groenke & Paulus, 2007).

To enhance student well-being, meaning-making and learning, we find it necessary to add dialogism to inquiry-based science teaching. We therefore use a sociocultural perspective on learning as meaning-making and mediated actions (see e.g. Wertsch, 1991). Fundamental to dialogic inquiry is an open atmosphere in the classroom, where feedback from teachers and other students is essential. If someone actively responds to an answer, a question or a statement, the situation of dialog will provide understanding to a much greater extent than when one-way communication is the predominant form of discourse (Bakhtin, 1981, 1986). Dialogic inquiry also requires that students can participate in a dialog with the subject matter, which can be done usefully in writing (Wellington & Osborne, 2001).

Episodes of dialogic inquiry are also situations in which learners and teachers explore ideas together that have not necessarily been planned for the lesson but are initiated by the direction taken by the dialog during classroom discussions (Wells & Ball, 2008). Dialogic inquiry is always based on talk about an interesting science issue that may be solved or otherwise can lead to an experimental investigation (Ash, 2003). Dialogic inquiry thus contributes to making hypotheses grounded on deep reasoning and observations, instead of starting an empirical investigation based on a pure guess. For example, many issues regarding sustainable development have no simple answers and need to be discussed in class, to evoke interest in urgent solutions and to cultivate responsible citizenship for the twenty-first century (see Constantinou, Tsivitanidou, & Rybska, 2018, this volume, p. 23–25).

Finally, we use dialogic inquiry as a framework in which we combine verbal communicative approaches (Mortimer & Scott, 2003) and writing (Martin, 1993) with learning for sustainable development (see UNESCO, 2011). These three approaches emphasise how the students and the teachers make meaning with the subject matter in focus and discuss topics relevant for students in their citizenship and as potential natural scientists. We claim that a teacher professional development programme, highlighting the importance of talking science and writing science, can help teachers to create this dialogic atmosphere in their classrooms (Osborne, 2006).

2 Background

In this section we will present the framework of dialogic inquiry behind the TPDP with the three parts mentioned above: communicative approaches, writing in dialog as a communicative tool and learning for sustainable development. Education for sustainable development has an interesting subject content, where verbal communicative approaches evoke societal discussions. It has also an interdisciplinary character as well as importance for our future.

2.1 Communicative Approaches

To create a classroom where dialog is essential, it is important that the teachers are aware of how to use different verbal communicative approaches (Mortimer & Scott, 2003). Thus the teacher gets the opportunity to meet the students in an honest dialogical-interactive talk, which can result in student contributions that develop the discussion. Mortimer and Scott (2003) made a didactic model for classroom communication in two dimensions, with four categories of communicative approaches: interactive/noninteractive and authoritative/dialogic (see Table 1).

Table 1 The teachers' different verbal communicative approaches (Mortimer & Scott, 2003)

Communicative approaches	Interactive	Noninteractive
Focus on science view Authoritative	The teacher introduces a topic by using questions and pupils' short answers	The teacher "lectures", and the pupils are supposed to listen

Taking account of pupils' understanding
Dialogic or nonauthoritative
	The teacher is
	  Probing
	  Elaborating
	  Supporting the students to give long comments and narratives of their experiences of a topic
		The teacher reviews the lesson and gives a summary of ideas that have been discussed or the alternative viewpoints that have to be considered for the next step

Interactive communication can either be authoritative (interactive-authoritative), such as when the teacher wants to initiate a short question and answer sequence, or dialogic (interactive-dialogic) when the teacher invites students' ideas and reflections in a more explorative way. Noninteractive-authoritative communication takes place when the teacher lectures and gives information about a topic, limiting student participation to listening. There is also a dialogic form of noninteraction. When the teacher reviews and summarises the students' ideas and contributions in the talk, the communicative approach is noninteractive but still dialogic, due to the content of student contributions (noninteractive-dialogic). Scott, Mortimer and Orlando (2006) argued, in a study on how secondary students make meaning of heat and temperature in a Brazilian science classroom, that there is a natural tension between the four different communicative approaches and that one can follow "authoritativeness acting as a seed for dialogicity and vice versa" (p. 605). In a dialogic science classroom, the manner of teaching means, for example, presenting scientific ideas, helping the students to understand and making science content the students' own and then also handing over the use of the content to the students themselves (Mortimer & Scott, 2003).

Alexander (2004, p. 28) stated how dialogic teaching follows five principles:

Collective;…teachers and pupils address learning tools together…
Reciprocal:…they share ideas, and consider alternative viewpoints
Supportive:…they articulate their ideas without fear
Cumulative:…they link their ideas into a chain of thinking …
Purposeful:…teachers have particular educational goals in view

Dialogic teaching and dialogic inquiry also include student-to-student interactions, for example, group discussions during practical work. Barnes and Todds (1995) discovered children's different varieties of talk: exploratory, when they talk to each other in small groups and together arrive at a solution to a given question, and as a final draft talk, when they turn to the teacher and sum up what they have done so far. We find exploratory talk very useful for the learners, and it is of importance that teachers working with inquiry-based science teaching learn to recognise students' use of exploratory talk. You can see exploratory talk (Barnes & Todd,

1995) as an opportunity for students to maintain a space of reflection with imaginative talk and to talk in a way that facilitates the emergence of creative solutions to problems (Wegerif, 2008).

2.2 Writing as a Communicative Tool

Dialog – often used synonymously with *conversation* or *talk* and primarily connected to the works of Bakhtin (1981) – is also used by others: Mercer (1995); Mercer, Wegerif and Dawes (1999); Wells (1999); and Wegerif (2008). In Bakhtin's theories (1981, 1986), dialog is a key concept, without distinction between written and oral texts. There are differences though between oral and written language use. In an oral dialog, the demands for clarity in communication are not pronounced, as the listener has the opportunity to ask questions if something is unclear. In writing there are other conditions. The writer does not always know who the reader will be or when, and under what conditions, the text will be read, which makes great demands on clarity. Writing makes students sharpen their thinking and use of appropriate concepts in an intelligent manner, which makes writing a unique tool in the learning process, a cognitively activating form of learning and an important part of scientific literacy (see Constantinou et al., 2018, this volume, pp. 20–21). This goes for both first and second language learners. For students with dyslectic problems, certain arrangements might be necessary. However, within the frames of this book, it is not possible to further develop this issue. Material gathered during the first year of S-TEAM showed, however, that writing was not a common element in instruction, and there were lessons that we observed where not a single word was written, neither by the teacher nor by the students (Norberg Brorsson, Engström & Enghag, 2014). Unfortunately, teachers often use writing in a one-sided way. This means that writing is only used to report something, usually to the teacher as the sole reader, or to provide one-word answers to textbook questions, not to create a dialog that affects thinking.

When studying science and other content subjects, the students have to learn certain concepts and certain ways of reasoning, embedded in an academic language that sometimes causes difficulties. Vygotsky (1978) distinguishes between *everyday language* and *academic language*. Students use everyday language when dealing with more or less practical and private issues. The decontextualised, abstract and logical academic language is, on the other hand, primarily developed at school and provides a language for scientific concepts and areas that are often not part of the students' everyday lives. Students have to master the academic language to be successful at school, and in this too, writing is an excellent tool.

Writing, reading and talking are firstly important tools for learning and, secondly, necessary competencies for university studies and jobs in the twenty-first century. In international writing research, there are two main directions corresponding to these two functions of writing. One direction, writing across the curriculum (WAC), stresses writing-to-learn strategies, which can be used in all subjects. (See for the early theories, e.g. Britton, 1970. One representative of today's researchers within this field is, e.g. Thaiss, 2012.) Spontaneous writing and writing to investigate can help the students "make the content of the subject their own". This kind of writing can be used in various ways in short mini-writing tasks such as: What do you know about…? What is your opinion of…? What solution do you think is best? Why? These texts primarily serve as the students' thinking tools, or as a basis for discussion and dialog in groups or in class, and should therefore not be marked by the teacher. The other direction, writing in the disciplines (WID), stresses the terminology of each subject, its linguistic style and the requirements of subject-specific genres, for example, the lab report in natural science (Halliday, 1978; Martin, 1993). In this kind of writing, the student's text has a reader, which makes writing more demanding than writing for his/her own thinking. To become a member of the discourse community of science and participate in its dialog, the students must have knowledge of the typical genres of science. In all the activities mentioned in Fig. 2 (Introduction: What Is Inquiry-Based Science Teaching and Learning?, p. 8), elements of writing can be included, be it writing-to-learn tasks or writing of genres. In the curricula for the compulsory Swedish school, including those of the natural sciences, the use of cognitively demanding genres and tasks, such as making generalisations, arguing, evaluating, analysing and drawing conclusions, is highlighted. In the curricula there are demands on both oral and written works. Although the theories behind WAC and WID are several decades old, they are still just as important in providing writing opportunities for the training of disciplinary, science discourses, which is shown also in the vital writing research of today.

2.3 Learning for Sustainable Development

As mentioned above, in the curricula, there are demands on both oral and written works based on science. When students make their inquiry, they ought to deal with authentic problems, ask questions and find solutions that they argue for, both in oral and written dialog. However, a common teacher view is that inquiry is a linear process from hypothesis to result without arguing and discussing (Davies, Petish & Smithey, 2006). But issues of sustainable development are well suited for dialogic inquiry, and in this section, we make clear the reasons for this.

In Chapter "Introduction: What Is Inquiry-Based Science Teaching and Learning?" (p. 9), it is said that there is a "need for citizens to be able to debate pressing socio-scientific issues from an informed position". One such pressing issue is sustainable development. Swedish school science embraces learning for sustainable development, where a dialogic inquiry-based approach gives the students the necessary knowledge in science. Learning for sustainable development is seen as a teaching approach (Sellgren, 2007), using a holistic view and a wide range of participatory, problem-solving methods tailored to the students. In education for sustainable development, this approach should also make use of, for example, discussion, illustration of concepts and opinions, exercises, experiments, modelling, role-playing, information and communication technology (ICT), case studies, excursions and outdoor learning, analysis of best practices and problem-solving (UNECE, 2005, p. 17) in addition. The students' personal experiences, science knowledge, learning processes, collaboration and acceptance of responsibility are central. It is appropriate to initiate conversations and writing about the limits of ecological settings (water cycle, energy flows, etc.) and then to touch on human needs, cultural difference, ethical issues and the technical strategies and solutions needed to solve future problems (Engström, 2008; Liepina & Jutvik, 2009). In short, the teaching approaches characteristic for sustainable development correspond to a large degree with those for dialogic inquiry.

3 Aims and Questions

With the TPDP, we intended to introduce dialogic inquiry, and in this chapter, we describe and analyse teachers' reflections on the impact of dialogic inquiry on their science teaching and their ability to analyse classroom situations with instruments regarding verbal communicative approaches, writing in dialog and learning for sustainable development.

We focus on the following questions regarding this research-informed TPDP:

- How do teachers describe their teaching situation in the science classroom?
- How do the teachers discuss their experiences of the research-informed parts of verbal and/or written communicative approaches and the use of topics from sustainable development?

4 The Training Module: Design and Instruments

In this section we describe the design of the TPD, the instrument developed for teachers to analyse video clips of earlier teaching and of their own teaching and the subject content.

4.1 The Design

Based on the ideas presented above, a training package for dialogic IBST/L was developed during *the first year* of the S-TEAM project, in collaboration with five teachers from two compulsory schools. Among the schools contacted by us, those two schools and the five teachers volunteered to take part in the project. Two teachers were educated for Swedish primary school, and the other three were science educators. The data were gathered in the form of video clips and lesson plans from different content areas – energy, electricity and the water cycle – in grades 5 and 6. The video clips were used as web-based resources, for the participants in the second year.

The pilot TPDP was given during the second year of the S-TEAM project in which video recordings and results from the first-year data formed important study material. Seven teachers volunteered to attend the pilot TPDP that included six meetings. These teachers were qualified teachers of technology and science for grades 4–6. None of them took part in the project during the first year. The design was built on the idea to implement a TPDP, which would give teachers the theoretical background to dialogic inquiry and to make analyses of video clips from others' as well as from their own teaching, and support them in developing their teaching and lesson planning. In the TPDP, the teachers reflected on different aspects of dialogic inquiry and read relevant literature on research (Alexander, 2004; Mortimer & Scott, 2003; Wellington & Osborne, 2001, and others). Between the meetings, the teachers analysed and developed their teaching by using qualitative video analyses designed by us (see Appendix 1, Tables 2 and 3) and by being observed by a colleague. The six meetings had different foci: (1) teachers' presentations and SWOT analysis of experiences, (2) analysing lesson content from video clips, (3) analysis and use of communicative approaches, (4) analysis and use of writing, (5) teaching for sustainable development, (6) results from analysing their own teaching and (7) reflections on the TPDP and the literature. The design of the first and second year of the project as well as of the pilot TPDP is shown in Fig. 1.

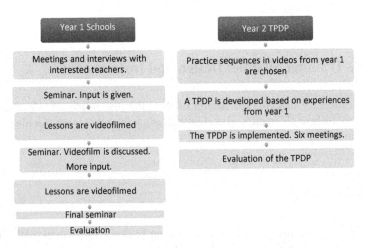

Fig. 1 The design of the first year in school and second year with the TPDP

5 Findings: The Teachers' Experiences and Reflections

5.1 The SWOT Analysis

A SWOT analysis is a common, qualitative tool admitting reflection and suitable as a starting point of discussions. The participants therefore filled in a matrix with a SWOT analysis where they mapped their strengths, weaknesses, opportunities and threats in teaching. The analysis was made individually, but we discussed the results in group, and we collected the SWOT analyses afterwards. From the SWOT analysis, we found that the teachers described themselves as strong leaders with resources to handle management in the classroom, with interest in experimental work and with a desire to develop their scholarship. They mentioned as strengths that they "liked teaching", "respected their students" and can "understand and see the students' different abilities and strengths". They found it troublesome, however, to "get everyone involved in the talks", and they feared "the lessons to be unstructured". Their own "subject knowledge was not deep enough", and they would like to be more qualified in formative assessment. They have good opportunities, with fine "resources such as smartboards, equipment" and "access to money" for activities. They experience good cooperation between primary and secondary education. The threats mentioned are "other things you need to do during these 40 minutes than to teach" that colleagues often leave school due to organisational changes and that "students have varying abilities to write". The considerable difference between what "I want to teach and what the textbook says" together with "too little time for planning" made teaching hard. They also found themselves lonely and sometimes driven by routine in their lesson planning. The teacher comment on textbooks is interesting. The impact of textbooks is often considerable among both teachers and students, although the teachers are free to decide the content of their instruction as long as it is in accordance with the Swedish national curriculum.

5.2 The Teachers' Observations of Use of Time and Communicative Approaches

As mentioned earlier, we had developed an instrument to record the use of time spent on different lesson activities. The teachers, in pairs, analysed and reflected on two video clips from the instruction during the first year of the project, based on topics in the curriculum: (1) a lesson on electricity in grade 6, when the class was building and using a charging indicator and using this to test how different materials conduct electricity, and (2) a lesson on the water cycle in grade 5.

A summary of the teachers' analyses showed that five categories of lesson activities were identified and quantified. In the course of doing this, a discussion of what were appropriate and well-functioning elements took place, and a common view was that too much time was spent on giving instructions, which made the students impatient and anxious to start the measurements: "Students are distracted by the instruction". A reflection made by the participants was that it is important that teachers express themselves clearly.

The teachers also found that the communication/interaction was characterised by: "A lot of monologue from the teacher, [only a] small part of the lesson was a dialog between teachers and students. Students (numerous) spoke during the experiment". They also addressed teacher talk: "The teacher asked questions, the students raised their hands and answered". Another anecdotal observation was about students' writing of hypotheses: "They write hypotheses before the hands-on experiments begin".

The teachers had no problem in seeing and addressing how the lesson activities had an impact on the learning situation for the students and how aspects of the dialogic classroom and dialogic inquiry were always there as an opportunity for the teacher to enter into, if he/she had enough awareness and had got a clear idea of students' questions and comments. The observations mentioned in the paragraphs above concern to a great degree activities that are linguistically conditioned and where teachers' knowledge could have made a difference. The teachers also reflected on three aspects as follows:

1. The teachers' ideas about how to manage dialogic inquiry: *Before you begin a section, brainstorm, preferably in writing, to capture what the students already know or what they think about. Build on the students' prior knowledge when testing different materials.*
2. How the teachers best can use their skills in the classroom: *Dare to let the lessons skid off the track even if that had not been the thought from the beginning. Seize the opportunity when it occurs, and latch on to subjects that students are taking up and to the students' spontaneous questions.*
3. How students can be encouraged to talk about their own experiences in class: *By talking and by showing that their opinions and reflections are important. Show that you respect all student questions.*

As mentioned above the teachers made analyses of their own lessons, and in the following, three extracts are given:

> By analysing the time distribution of activities, I have learned a way to evaluate my lessons and gained insights into how I can develop things in a lesson so that students become more active and learn better. In the literature there are many good examples of activities that I can use to vary my teaching.

> I have become more aware of how to ask open questions that all students can reflect on at any level. Another aspect the analysis made me aware of was that the students needed more group discussions – it becomes too much and too fast whole class talk. This was also commented on by my observer. When I had the next group, I put in more discussions in small groups before whole class talk and noticed that this was a positive [way] to make students participate.

One teacher reflected on his use of different communicative approaches after having been observed by a colleague, who was not participating in the TPDP:

I have long worked to develop the ability to use whole class talk: invite students into discussions, make suggestions, give reasons for their opinions, keep an open climate and so on. The dialogic attitude has given me   deeper awareness and thus a better framework for how discussions can be managed and implemented. I am constantly exploring opportunities to embrace dialog in education and allow students' ideas to interact to a greater degree. Moreover, I have become more aware of how teachers can navigate between the scientific perspective and everyday thinking by switching between more dialogic and more authoritative communication.

We observed that the experienced teachers could more easily use and analyse communicative approaches in teaching. They were confident because they had deep subject knowledge. Teachers with less subject knowledge, however, took the opportunity to develop new teaching materials from examples in the literature and tested them, often successfully, in the classroom. When it came to activities in the classroom, we discussed opportunities for enhancing the amount of subject-related activities and to minimise the time spent on management and control – as a way to optimise both student and teacher ownership and learning. These teachers supported each other during the course. Support from colleagues is important, and we observed that the participating teachers supported each other during the course (Appendix 2, Table 4).

5.3   The Teachers' Analysis of the Students' Use of Writing in Science

The teachers made an analysis of the students' use of writing in science and of language use generally, from video clips in the course and from their own teaching between the course meetings. An example of such an analysis is shown in Table 5, Appendix 2, where a teacher describes a whole teaching sequence, not just a single lesson, to show the linguistic use in his instruction during a longer period of time. The same teacher made some interesting reflections on his analysis:

The table shows that the students work orally to a high degree. In the future I must more consciously give the students opportunities to use more genres both in reading and writing. I must consciously plan for more use of multimodal texts. The students must be taught to take, give and write instructions. I will use … (mini-writing and writing of logbooks) more in the future. I would like to develop my competence to formulate and use open writing activities: "What do you think when …?" "How would you like to explain …?" I will let the students learn to describe in writing the objective qualities of substances, to train systematically in writing realisations, hypotheses, observations, and conclusions by the use of scaffolding. I will let the students read texts more systematically on problems we are working on.

The table of analysis shows clearly what genres the students meet, and in the future I will use it as a frame for planning my instruction to see to it that all aspects of reading, writing, and talk will be used and trained in instruction to a higher degree. I have compared my analysis to the curriculum in chemistry, and I can see that there are more genres that must be introduced in order to work according to the intentions of the curriculum.

The reflections above also apply to a high degree to the other teachers of the TPDT. From their analyses, it may be concluded that the use of writing in their instruction had increased during the course of the TPDT. For most of them, writing had up till then not been a great issue. Some of them had, however, given their students extensive writing tasks, mostly as part of reporting, but not for other dialogic purposes. According to the teachers' reflections, they used writing in instruction in an unconscious way, because they lacked the knowledge of how to use it as an integrated part of teaching and learning science. If the teacher, for example, asks the students to write down definitions, their suggestions of solutions or the questions they still have at the end of the lesson, then writing can be used as a starting point for discussions where more students than is usually the case can contribute because they have had time to think and reflect and the teacher gets information of the students' knowledge and thoughts. Apart from this, the students get an opportunity to use the subject-related academic language in writing. The new national curricula, implemented during the TPDP, are more linguistically oriented than the former ones, and tasks where writing is required are highlighted in science as well. This worried the teachers as they felt that they lacked the necessary tools. The TPDP had, however, started a process where writing, along with reading and talking, would get greater focus in their instruction and thus support the development of their students' scientific literacy. Once the teachers become aware of the benefits of writing, they will use it in a more reflected and systematic way.

5.4   Teachers' Choice of Topics Regarding Learning for Sustainable Development

As an example of how one of the teachers worked with dialogic inquiry in science teaching with the aim of making the students learn for sustainable development, a sequence of the teacher's report of his chemistry teaching is highlighted. In the sequence, aspects of sustainable development are implicit rather than explicit. The students worked with different substances on a tray (Hallgren, 2010), presented by the teacher. In a dialog, the teacher introduced central concepts to help the students' understanding of many of the processes that cause future problems. They were given challenges to discuss and to write conclusions about environmental problems in relation to chemical reactions. Below is an example of subject content-related results that emerged in the students' writing of reports and in conclusions from the classroom dialog:

- Greenhouse gas and greenhouse effect: carbon dioxide is formed in a chemical reaction between carbon and oxygen.
- Pure metals must be produced and take a lot of energy; recycling is often a better option.
- The reaction between sulphur and oxygen contributes to acid precipitation.

- Discussion of mass conservation in a chemical reaction and new substances (e.g. gases) produced and disseminated.
- Soil chemical composition and role as an environmental factor in the ecosystem.
- Prerequisites for the understanding of cycles.

The teacher introduced different aspects in the dialog with the aim of getting a broader understanding, and not only a focus on the chemical reactions and reflecting on the hypothesis. Moreover, the students got a richer and more concrete experience of substances and the processes of importance in discussions of sustainable development. In this way dialogic inquiry enhances students' learning and creates interest and engagement.

6 Discussion and Conclusions

We start this last section by discussing the teachers' experiences of their teaching and progression from the TPD.

6.1 Teachers' Descriptions of their Teaching Situation in the Science Classroom

The teachers gave a picture of being strong leaders in the classroom, with great interest in the students' well-being and learning outcomes but also with a fear of not being able to cope with new challenges and pedagogical demands, which is also reported in the literature (see, e.g. Guskey, 2010). They themselves pointed out whether they were confident with their subject knowledge or not, and this varied in the group. As already mentioned in Chapter "Introduction: What Is Inquiry-Based Science Teaching and Learning?", the teacher's professional competence is a key issue and a prerequisite for successful IBSL/T. The teachers need not only good subject knowledge; they also need fantasy to create a good teaching and learning environment, which sometimes is very important for the learners.

6.2 Teachers' Experiences of the Research-Based Parts of Communicative Approaches and Writing in Dialog

We observed that the more confident the teachers were in their specific subject, the easier they had to fulfil the analyses we asked them to maintain during the programme. The teachers were informed that we had analysed these video clips before in the course. Regarding communicative approaches, they observed that in the beginning, teachers in the video clips mostly used a noninteractive/authoritative lecturing style, which made the students restless. When an interactive/authoritative style was used, the students gave short answers and – by raising their hands – a sign that the atmosphere was not very open and dialogic. An interesting teacher observation was that the students had an opportunity to reason themselves during the small-group work, which in a way excluded the teacher from the students' learning process. In these situations, the teacher has to accept the loss of control of the students' discussions for a while. The teachers' first ideas regarding writing concerned the writing of hypotheses, which was used when they took notes to report their practical investigations. During the TPDP various writing-to-learn strategies seemed to attract great interest, perhaps because they are easily integrated in the instruction. The use of different genres, on the other hand, requires the teacher's knowledge of genre-specific features, something the teachers said they needed to manage, but it had not been part of their teacher education. While waiting for further teacher professional development (TPD) programmes where also subject-specific language issues are focused, they will have to manage their instruction partly without the positive and important effects writing in science can have on students' scientific development.

6.3   Reflections and Conclusions

The Vygotsky-inspired ideas for our project have been essential for the teachers' experience of how important the dialog is for students' activities in the classroom. The design of the teacher professional development programme gave us also some valuable experiences. It is evident that the video clips from the first year science instruction played an important role in designing the course during the second year, and they inspired the course participants in planning and analysing their own instruction – how teachers teach teachers.

The fact that the video clips were possible to look at in a web centre facilitated the use of them. The video clips were possible to look at in a web centre. The information and seminars based on analysing video clips from other teachers' efforts were most helpful in conveying the theoretical ideas we draw upon, the Vygotskian ideas (1978) that:

> language not only function as a mediator of social activity, by enabling participants to plan, coordinate and review their actions through external speech; in addition, as a medium in which those activities are symbolically represented, it also provides the tool that mediates the associated mental activities in the internal discourse of inner speech (Wells, 1999, p. 7).

Of course the ideas on verbal and written communication and dialog through specific subject knowledge are necessary in teaching, and they are through Mortimer and Scott, Osborne and Wellington and others also representative for the ideas of Vygotsky and Bakhtin.

Tasks for teachers to work with in between meetings, and the use of literature, guided the development of new teaching materials with new subject content, for example, with relation to sustainable development. The course inspired commitment among the teachers to science education research. The teachers also found an arena to tell and show each other their own good ideas. They discussed the value of dialog as a link to learning and the use of writing as a dialogic tool for teaching and learning. The development of writing in science, and other subjects, needs time and support, and the teachers must be confident in how to meet the students' needs. If not, the teaching will offer the students too few opportunities to practise and to develop their writing skills (Martin, 1993). In addition, writing tasks are often given without clear instructions and expectations, as well as without guidelines related to the structure and organisation of a specific text type. To make the students scientifically literate, the teacher has to specifically teach the students how to write the typical genres of science (Wellington & Osborne, 2001) as well as various writing-to-learn strategies. It is therefore of importance that science teachers are offered opportunities to develop their competence within the linguistic field as well.

We found that teachers who lacked confidence in the subject when studying the videos also had difficulties in analysing what kind of content (concepts, applications or socio-scientific issues) was in use during a lesson. However, they took the opportunity to develop new teaching materials from examples in the literature and tested them successfully in the classroom. Teachers with solid subject knowledge, on the other hand, did not face problems analysing the videos.

One example is how we have experienced that inquiry, such as an empirical investigation in the Swedish schools, often starts with students guessing a hypothesis. We argue that dialogic inquiry is a way to generate well-grounded questions before the start of an empirical inquiry and enhance students' engagement, both in the actual inquiry project and in environmental issues or other sustainability aspects in general. Thereby the chain between the steps in the investigation will be much more important and visible to the pupils. Guessing is an important part of abduction (Tschaepe, 2014), which is a logical process that demands critical thinking and deliberation.

Formulating the question into a hypothesis which can be tested with an approachable method stands out as the most crucial and important factor in IBST/L, and dialogic inquiry makes that step significant.

One way to include teachers as participants and with ownership of the teaching process of their professional development (Adey, 2004) is to ask for good experiences and use these as a starting point to collaboratively develop teaching, with input from research-based ideas and results.

To have long-term effects and to form a basis for new efforts, a teacher professional development programme must give ownership to the teachers and be based on theories and research, and not be a quick fix of tips or examples out of context. The teachers need regular opportunities to discuss the meaning of dialogic inquiry and to realise the learning potentials in the use of writing, both mini-writing tasks and in different genres. As a foundation for all this, sufficient subject knowledge is crucial. If this is not the case, the risk of the teacher not daring to develop a dialogic classroom is obvious. The S-TEAM project was too short to give us any proof in terms of students' enhanced learning and interest, but a TPDP of the design described above gives the teachers tools to cope with many of the problems they face in their classrooms.

Appendix 1

Table 2 Instruments for the teacher to measure how they used their lesson time on different lesson activities and whether essential content from areas we focused on was addressed

A	Lesson analysis for Name... Date.................
	Notes...
	School syllabus and curriculum

School subject - Education for sustainable development

Instructions/ Management	Science Content	Writing in science	"Just-Talking"	Democratic communication		Value-laden training	Out-of-school-connections	Socio-scientific issues	Activity
				Small-groups	Class				
TIME									
1									
2									
etc									

Table 3 Instruments for the teacher to measure how they used different communicative approaches during science discussions

A	School subject – Education for sustainable development							School syllabus and curriculum		
Communicative-approaches -Interaction	Instructions/ Management	Science Content	Writing in science	"Just Talking"	Democratic communication		Value-laden training	Out-of-school-connections	Socio-scientific issues	Activities
					Groups	Class				
Non-interactive										
Interactive										
Monologic										
Dialogic										
Concepts										
Applications										
Context										
WA										
WD										
WU										
Activities										
TOTAL										

Appendix 2

Table 4 The use of reading, oral discussion and writing of different genres

What are my students supposed to do?	Narrative	Recount	Description	Argumentative text (oral or written)	Explanation	Instruction
Reading						
Talking						
Writing						

Adaptation of Holmberg (2010)

Table 5 The use of reading, discussion and writing of different genres

What are my students supposed to do?	Narrative	Recount	Description	Argumentative text (oral or written)	Explanation	Instruction
Reading			Definitions of concepts. To read and use tables, encyclopaedia		Passages in a textbook, e.g. on chemical reactions	Laboratory instructions
Talking		Recount in their own words, repetition of definitions of concepts	Describe substances on a tray. The use of concepts to describe and explain occurrences and experiments	Motivate and discuss classifications, make and motivate hypotheses, discussions on connections	Explain connections between various concepts, explain what happens in experiments/demonstrations	When making experiments
Writing		Mini-writing about concepts in their own words/ definitions	Describe chemical reactions before, during and after an experiment. Describe substances on a tray. Make a map of connections between substances	Make hypothesis and motivate	Explain results in their own words	

Adaptation of Holmberg (2010)

References

Adey, P. (2004). *The professional development of teachers: Practice and theory.* Dortdrecht/Boston/London: Kluwer Academic Publishers.

Alexander, R. J. (Ed.). (2004). *Towards dialogic teaching: Rethinking classroom talk.* London: Dialogos.

Ash, D. (2003). Dialogic inquiry in life science conversations of family groups in a museum. *Journal of Research in Science Teaching, 40*(2), 138–162.

Bakhtin, M. (1981). Discourse in the novel. In C. Emerson & M. Holquist (Eds.), *The dialogic imagination. Four essays.* Austin, TX: University of Texas Press.

Bakhtin, M. (1986). The problem of speech genres. In C. Emerson & M. Holquist (Eds.), *Speech genres and other late essays.* Austin, TX: University of Texas Press.

Barnes, D., & Todd, F. (1995). *Communication and learning revisited: Making meaning through talk.* Portsmouth, NH: Boynton/Cook Publishing.

Britton, J. (1970). *Language and learning.* Harmondsworth, Middlesex: Penguin Books.

Brorsson, B., Engström, S., & Enghag, M. (2014). Muntlig kommunikation under en lektion om energikällor i årskurs 5 [Oral communication during a lesson on energy sources in year 5]. *NorDiNa, 10*(1), 46–62.

Constantinou, C. P., Tsivitanidou, O., & Rybska, E. (2018). Introduction: What is inquiry-based science teaching and learning? In O.E. Tsivitanidou, P. Gray, & E. Rybska (Eds.), *Professional development for inquiry-based science teaching and learning* (Contributions from Science Education Research 5). Cham, Switzerland: Springer.

Davis, A. M., Petish, D., & Smithey, J. (2006). Challenges new science teachers face. *Review of Educational Research, 76*(4), 607–651.

Edelson, D. C., Gordin, D. N., & Pea, R. D. (1999). Addressing the challenges of inquiry-based learning through technology and curriculum design. *Journal of the Learning Sciences, 8*(3–4), 391–450. https://doi.org/10.1080/10508406.1999.9672075.

Engström, S. (2008). *Fysiken spelar roll! Undervisning om hållbara energisystem, fokus på gymnasiekursen fysik A* [Physics plays a role! Instruction on sustainable energy systems, focus on the upper secondary course physics A]. Licentiate Thesis. Eskilstuna: Mälardalen University Press.

Groenke, S. L., & Paulus, T. (2007). The role of teacher questioning in promoting dialogic literary inquiry in computer-mediated communication. *Journal of Research on Technology in Education, 40*(2), 141–164.

Guisti, B. M. (2008). *Comparison of guided and open inquiry instruction in a high school physics classroom.* Brigham Young: School of Technology Brigham Young University.

Guskey, T. R. (2010). Professional development and teacher change. *Teachers and Teaching, 8*(3), 381–391. https://doi.org/10.1080/135406002100000512.

Gyllenpalm, J., Wickström, P.-O., & Holmberg, S.-O. (2010). Nordic. *Studies in Science Education, 6*(1), 44–60.

Hallgren, L. (2010). *Kemibrickor i ständig utveckling* [Chemistry trays in continuous development] http://slidegur.com/doc/4901655/lisbeth-hallgrens-m%C3%A5lformulering

Halliday, M. A. K. (1978). *Language as social semiotic. The social interpretation of language and meaning.* London/New York/Melbourne/Auckland: Edward Arnold.

Holmberg, P. (2010). Text, språk och lärande. Introduktion till genrepedagogik [Text, language and learning. Introduction to genre pedagogy]. In M. Olofsson (Ed.), *Symposium 2009.* Genrer och funktionellt språk i teori och praktik (pp. 13–27). Stockholm: Stockholm universitets förlag.

Kouns, M. (2014). *Hur lärare utvecklar språkinriktad fysikundervisning: en designstudie på gymnasiet* [How teachers develop a linguistically oriented physics instruction: A design study in upper secondary school]. Malmö Studies in Educational Sciences No. 73, ISBN 978-91-7104-587-4. ISSN 1651-4513.

Liepina, I., & Jutvik, G. (2009). *Education for change. En handledning för undervisning och lärande för hållbar utveckling* [A manual for the teaching and learning of sustainable development]. Uppsala: Uppsala Universitet. Hämtad i mars, 2011 från www.balticuniv.uu.se

Linn, M. C., Davis, E. A., & Bell, P. (2004). *Internet environments for science education.* New Jersey: Lawrence Erlbaums Associated.

Martin, J. R. (1993). Literacy in Science: Learning to handle text as technology. In M. A. K. Halliday & J. Martin (Eds.), *Writing Science: literacy and discursive power* (pp. 166–202). Pittsburgh, PA: University of Pittsburg Press.

Mercer, N. (1995). *The guided construction of knowledge – Talk amongst teachers and learners.* Clevedon, Avon, UK: Multilingual Matters Ltd.

Mercer, N., Wegerif, R., & Dawes, L. (1999). Children's talk and the development of reasoning in the classroom. *British Educational Research Journal, 25*, 95–111.

Mortimer, E. F., & Scott, P. H. (2003). *Meaning making in secondary science classrooms.* Buckingham, UK: Open University Press.

Osborne, J. (2006). *Towards a science education for all: The role of ideas, evidence and argument.* In Proceedings of the ACER conference: Boosting science learning – What will it take? Camberwell, VIC: ACER.

Ottander, C., & Ekborg, M. (2012). Students' experience of working with socioscientific issues – A quantitative study in secondary school. *Research in Science Education, 42*(6), 1147–1163. https://doi.org/10.1007/s11165-011-9238-1.

Scott, P. H., Mortimer, E. F., & Aguiar, O. G. (2006). The tension between authoritative and dialogic discourse: A fundamental characteristic of meaning making interactions in high school science lessons. *Science Education, 90*(4), 605–631.

Sellgren, G. (Ed.). (2007). *Lärande på hållbar väg* [Learning the sustainbable way] Solna: Världsnaturfonden WWF.

Swedish National School Agency. (2012). *Greppa språket – Ämnesdidaktiska perspektiv på fler språklighet* [Grab the language! Didactic perspective on multilingualism]. http://www.skolverket.se/publikationer?id=2573

Thaiss, C. (2012). Origins, aims, and uses of writing programs worldwide: Profiles of academic writing in many places. In *Writing Programs Worldwide: Profiles of Academic Writing in Many Places* (pp. 5–22). Perspectives on Writing. Fort Collins.

Tschaepe, M. (2014). *Guessing and scientific discovery: Hypothesis-generation as a logical process.* Retrieved 2016-02-06 http://www.hopos2014.ugent.be/node/351.

UNECE. (2005). *United Nations Economic Commission for Europe.* Strategy for education for sustainable development. CEP/AC.13/2005/3/Rev.1. Retrieved June 2009 at http://www.unece.org/env/documents/2005/cep/ac.13/cep.ac.13.2005.3.rev.1.e.pdf

UNESCO. (2011). *Education for Sustainable Development (ESD).* Retrieved July 2011 at http://www.unesco.org/new/en/education/themes/leading-the-international-agenda/education-for-sustainable-development/

Vygotsky, L. (1978). *Mind in society. The development of higher psychological processes.* Cambridge. MA: Harvard University Press.

Wegerif, R. (2008). *Dialogic, education and technology. Expanding the space of learning.* New York: Springer.

Wellington, J., & Osborne, J. (2001). *Language and literacy in science education.* Buckingham, Philadelphia: Open University Press.

Wells, G. (1999). *Dialogic inquiry. Towards a sociocultural practice and theory of education.* Cambridge, UK: Cambridge University Press.

Wells, G., & Ball, T. (2008). Exploratory talk and dialogic inquiry. *Exploring Talk in School, 13*(2), 167–184.

Wertsch, J. V. (1991). *Voices of the mind: A sociocultural approach to mediated action.* Cambridge, MA: Harvard Univerisity Press.

Designing Teacher Education and Professional Development Activities for Science Learning

Andrée Tiberghien, Zeynab Badreddine, and David Cross

1 Introduction

Inquiry-based teaching is complex and includes many different aspects, from designing relevant problem-based activities to developing student autonomy and discursive argumentation between peers and in the whole classroom. This teaching orientation should encourage students to learn science: its content, its epistemology (Nature of Science), its value and its relevance for the study of societal questions.

This chapter is focused on some basic components of inquiry-based science teaching (IBST), designed to develop students' autonomy, in relation to some general aspects of learning science. The chapter particularly deals with what we call "students' intellectual autonomy" in a scientific domain. This is not only a form of autonomy related to the actions they decide to carry out for experimental activities, but it is also the autonomy to construct new knowledge, which in turn implies that they develop a responsibility vis-à-vis knowledge.

To develop this autonomy, we take a theoretical approach for which the goal of teaching is to develop understanding of content, procedural and epistemic knowledge and that focuses on the teacher and students joint actions to achieve this goal. This choice of actions implies a holistic perspective in the sense that the relationships between knowledge, teaching and learning are conceptualized. This allows us to consider IBST as a basic choice to teach science, like it is presented in chapter. "Introduction: What Is Inquiry-Based Science Teaching and Learning?", since it is related on one hand to opportunities to learners for achieving a better understanding of science concepts, principles and phenomena and on the other hand to learner's metacognition like process skills, critical thinking, decision making, etc.

A. Tiberghien (✉) · Z. Badreddine · D. Cross
UMR ICAR, CNRS, University of Lyon, Lyon, France
e-mail: andree.tiberghien@univ-lyon2.fr; zeynab_badreddine@yahoo.fr; david.cross@umontpellier.fr

© Springer International Publishing AG, part of Springer Nature 2018
O. E. Tsivitanidou et al. (eds.), *Professional Development for Inquiry-Based Science Teaching and Learning*, Contributions from Science Education Research 5, https://doi.org/10.1007/978-3-319-91406-0_13

The conceptual network used in this chapter called "theory of joint actions in didactics" guides us first to analyse the practices of a 10th grade physics classroom during a teaching sequence on mechanics. This analysis is focused on how the progression of knowledge and scientific practices are developed in the classroom in relation to the evolution of the respective students' and teacher's ownership of knowledge, which is a way to acquire intellectual autonomy. Let us note that here IBST is used for teaching across a typical science domain and not when teaching some specific science content or socio-scientific issues. This is particularly important to the extent that science domains such as classical mechanics are often taught in a "traditional way".

In the second part of the chapter, this framework also guides us to discuss the teachers' choices and actions to propose bases to design resources for teacher development.

2 Classroom Analysis

2.1 *Theoretical Framework*

In our theoretical framework, the classroom is approached from a didactic perspective in order to account for its practices. This framework is based on the theory of joint action in didactics. In this theory, the main object of study is the classroom, viewed as a community of practice where two joint actions are involved: teaching and learning (Sensevy, 2007). These two joint actions are based on communication between the teacher and students and between students. Due to the instructional goals given by society to school, knowledge is at stake in classroom communication. In most countries, this goal is made explicit through official texts including standards or an official curriculum. An important component of classroom communication is the reciprocal expectations that the teacher and the students may have; Brousseau (1997) called this the "didactic contract". This contract forms a system of norms or habits, some of which are generic and will be lasting and others, which are specific to current elements of knowledge, need to be redefined when new elements are introduced. For example, after the teacher has introduced the concept of force in a physics class, his/her expectations of the students' interpretations of material situations will be different from before, particularly concerning the justification of the interpretations. Another important component is "the milieu" that is the social and material components with which the students construct knowledge meaning. Thus, understanding classroom practices necessitates understanding the temporal evolution of the didactic contract and of the milieu, not only on the teaching or learning side but also on the side of joint teaching-learning actions.

In the frame of the didactic contract, two types of moment related to the status of certainty of knowledge are important. There are moments where the class group accepts that the ideas under discussion are only propositions and where the students recognize that they do not know this scientific knowledge; such moments are necessary to construct new knowledge. We call them "moments of epistemic uncertainty" (Tiberghien, Cross & Sensevy, 2014). The other type of moment, called "institutionalization", occurs when the teacher decides to tell the students that their activity has enabled them to construct knowledge that is legitimate in institutions outside the classroom (like scientific communities) and to make them take account of such knowledge in future actions (Sensevy, 2007). Even if the status of some elements of knowledge evolves during these two particular moments, a continuity of knowledge is necessary; thus relationships between these elements of knowledge are established.

The institutionalization does not imply that students have necessarily learnt this knowledge. We differentiate the student learning pathway called "the learning time" from the rhythm of introduction of new knowledge in the classroom called "the teaching time".

Note that we do not specifically focus here on the verbal temporal links that are made explicit in the teacher's discourse but rather on the temporalities of the teacher's action in constructing a didactic contract and a milieu (activity, classroom organization, etc.) in the classroom (Badreddine & Buty, 2011; Mercer, 2008).

2.2   Research Questions

Consequently, the following research questions deal with the teachers' actions associated with the introduction and progression of knowledge during a teaching sequence associated with the development of students' intellectual autonomy, a central component of IBST. These actions change some aspects of the didactic contract, but they are also dependent upon the contract already established in the previous sessions.

1. What kind of didactic contract favours the continuity of knowledge in the classroom between students' propositions and the knowledge that the teacher should introduce according to the official curriculum?
2. What are the actions that the teacher should carry out in a classroom in order to foster a didactic contract and a milieu which may enable students to acquire knowledge, and to develop intellectual autonomy, and more generally a scientific approach?

2.3   Main Components of the Evolution of an Element of Knowledge in the Classroom

The data used in this chapter were collected in the context of a research-based design project for teaching sequences (Tiberghien, Vince, & Gaidioz, 2009). The teacher followed a teaching sequence in mechanics at grade 10 elaborated in the context of this project. We succinctly present our analysis of the classroom practices.

For the previous study, two classrooms with different teachers were observed during the teaching of the topic "dynamics" (six sessions for a teacher and seven for the other one). All the sessions were videotaped with two cameras, one covered the teacher and part of the class, whilst the other one covered two students (the same students during the whole teaching sequence) and a part of the class (Malkoun, 2007; Tiberghien & Malkoun, 2010). The two students of each class were chosen by the teacher (at the request of the researchers) to select students with a middle or low level and who discuss with each other.

The conceptual structure of the sequence is based on epistemological choices regarding modelling, differentiation between concepts and objects/events in the material world. This choice leads the designers to use the word "force" only with its meaning as a physics concept and not with its everyday meaning. Therefore, the word "action" designates the event: an object acts upon another object. Thus, the notion of action is introduced first and then the concept of force, and finally the inertia principle is introduced. Let us note that the idea of object is already the results of a categorization which is not the same in physics and in everyday life. In physics, following the Newton law of universal attraction, any object (e.g. a book, a small stone, a hair, or the planet Earth) can be modelled as a point mass and thus belongs to the same category of material objects, whereas in everyday life, most of the time, an object is subject to manipulation (which is not the case of the Earth).

In order to discuss the teacher's action in the observed classroom, we present the evolution of the classroom during six sessions dealing with the introduction of dynamics in grade 10. This presentation is focused on a specific element of knowledge: the differentiation between the action of the ground and of the Earth. The difference between the actions of the Earth (the planet) and the ground (e.g. the solid surface of the Earth) is based on experimental considerations: they have opposite directions, and the effects of their action on objects are the reverse, the Earth attracts downwards, and the objects fall down, whereas the ground (or any support) prevents an object from falling down. Let us note that from the scientific argumentation perspective, not all elements of knowledge are associated with experimental evidence. This is the case for the first law of Newton (inertia principle), which is constructed, like the other physics principles, by scientists and is true until a series of experimental facts contradict it and are recognized as such by the scientific community (Valentin, 1983). Due to their different epistemological status, the learning pathways to acquire these elements of knowledge are also very likely to be different. In the case of the inertia principle, it is the first time that students have learnt a physics principle, and they have to acquire the way of thinking based on a principle, whereas they have already acquired elements of knowledge based on experimental facts, even if they are related to concepts. This chapter is focused on the first element of knowledge, the differentiation between the actions of the Earth and of the ground on an object (like a book).

Let us note that choosing a particular element of knowledge does not mean that it is "isolated" from other elements; on the contrary, we emphasize the importance

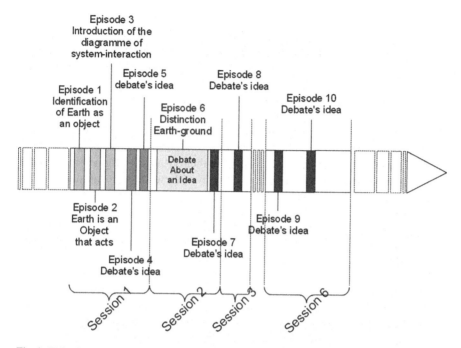

Fig. 1 Episodes where the element of knowledge, "the difference between the action of the ground and of the Earth", is involved in the timeline of the teaching sequence on mechanics (10th grade) (during sessions 4 and 5, this element is not involved)

of explicitly relating elements of knowledge to improve science understanding and learning (Roth et al., 2011).

We present the series of episodes in a timeline corresponding to the teaching sequence in the classroom (Fig. 1). The selection of episodes is based on a systematic analysis of the classroom discourse; when the students work in small groups, we take the discourse of the videotaped small group (two students) and the discussions of other small groups when the teacher intervenes.

- In episode 1 (session 1), the idea that the Earth is a material object emerges in small groups.
- In episode 2 (session 1), the teacher in whole class confirms that the Earth is a material object and that this element of knowledge is, therefore, certain from the physics perspective. Thus it becomes a public element of the physics classroom knowledge we can say that it is institutionalized.
- In episode 3 (session 1), the teacher introduces a formal representation of interactions called the diagram system-interactions where an ellipse represents an object, a full double arrow a contact action and a dotted double arrow a distant action (see, e.g. Appendix, Fig. 2). This associated knowledge is presented as not being open to question and is accepted as such by the students who use it rather easily in the following activities.

- Episode 4 (session 1) shows the emergence of the idea that the actions of the ground and of the Earth are not the same, proposed by a student working in a pair with another student during the activity where the students have to draw the diagram system-interactions showing all the objects that act on a table where there is an object (like a book).
- Episode 5 (session 1) also corresponds to work in a small group where the teacher helps students to clarify their ideas.
- Episode 6 (session 2) corresponds to a debate that takes place at the whole class level during the correction of the activity on which students work in episodes 4 and 5. During this debate, students present their ideas as being possible but not necessarily correct; now at the whole class level, these ideas are questioned. This questioning has evolved from private (small group) to public status (whole class). The debate ends with an intervention by the teacher who gives rational arguments. At this point, there is an institutionalization of this element of knowledge supported by these rational arguments (e.g. the ground or the soil prevents the table from sinking downward).
- In the next episode (episode 7, session 2), the teacher makes it explicit that this knowledge is now public and is recognized as a part of the physics knowledge of the class; in other words, she institutionalizes this knowledge.
- The following episodes (8 (session 3), 9 and 10 (session 6)) show that some students have not learnt these elements of knowledge after their institutionalization. In three sessions after session 6, several students, in small groups and in the whole class, are still having difficulties in using this difference between the ground and the Earth in different material situations.

3 Main Teaching Conditions to Establish Students' Intellectual Autonomy

On the basis of our theoretical framework and classroom analysis, we propose some main conditions, to establish a didactic contract and a milieu to enable students to learn science by developing intellectual autonomy. Four main conditions can be selected from our framework: sharing some common knowledge and meaningful vocabulary, managing moments of "epistemic uncertainty", institutionalizing the main elements of knowledge involved in the previous class activity and differentiating teaching and learning time.

3.1 Premise of Developing New Issues: Sharing Some Common Knowledge and Meaningful Vocabulary

Figure 1 shows that the first three episodes in session 1 are dealing with activities about the idea of action between objects and about learning or relearning that the Earth is an object. In everyday life, the notion of objects is limited to objects that can be handled.

In Newtonian mechanics, the law of attraction is relevant for the Earth or a book modelled in the same way; they are in the same category of objects. As introduced above, the categorization of objects in everyday life and in physics is not the same.

These episodes illustrate that students should learn some basic notions that are often considered as obvious, and are not made explicit in the official curriculum; they are, however, premises of classical notions or concepts presented in the official curriculum.

In terms of actions, this implies that the teacher, when preparing a teaching sequence, should be aware of this, should try to identify these basic notions and should design classroom activities allowing the students to construct or reconstruct these premises. This allows students to share the same elements of knowledge and therefore the same vocabulary with a shared meaning in the classroom. This necessity of sharing common knowledge to construct arguments and new ideas is also particularly important when problems in IBST come from everyday life or social situations, because the meaning of the words and expressions used to introduce the problem is not identical to those used to construct hypotheses and questions from a scientific point of view. The teacher should be aware of this and should be careful, when supporting discussions, that students understand each other. S/he can be enabled to design activities that allow the students to share basic common knowledge and an associated vocabulary. This sharing must be supported by a didactic contract, where the students are responsible for knowledge, in order to discuss and develop their argumentation in constructing new ideas.

3.2   Development of New Ideas with Students: Managing Moments of "Epistemic Uncertainty"

Episode 4 illustrates a moment where the students are aware that they do not know how to solve the problem but "play the game" to work on it and construct propositions. In this episode (Fig. 1), the students M and C are working together on the following question: draw the diagram system-interactions of a table on which an object (like a book) is set (see the right solution Fig. 2 part 2, Appendix). Before the point where the excerpt begins, the two students have agreed on their answer, i.e. that the object and the Earth are acting on the table (Fig. 2 part 1, Appendix); they have just had a short interaction with the teacher, and then they start to write their answer. However, one of the students stops writing and asks her peer whether or not the Earth and the ground are the same (see the extracts given in the Appendix, turns 1 and 5). This question emerges from the students' discussion in the group work situation, where they have to identify what is acting on the table and distinguish between distant and contact action. Here, as we explained before, the students are familiar with the notion of action, the type of questions and the diagram: they do not ask questions about how to draw it. The exercises and in particular the series of situations to analyse (before and after the situation with the table) and the use of the symbolic representations of the diagram system-interactions help students to raise questions about the difference between the ground and the Earth. Thus, this questioning emerges from knowledge as presented above.

This example illustrates that, through the didactic contract established in the classroom, the students are ready to construct an answer with justification but this answer does not have to be the correct one. It also shows that the teacher only helps the students to understand the situation. This is a moment of "epistemic uncertainty". The teacher expects the students to construct new propositions, and the students expect the teacher's help in understanding rather than in finding the correct answer. Usually, these reciprocal expectations slowly develop into habits in the class, when the teacher constructs them from the beginning of the academic year, but it can take several weeks or even months to develop the habit. This moment is possible because the activity and more globally "the milieu" are adapted; it involves a semiotic representation (the diagram Appendix, Fig. 2) and the notions of distant and contact actions that are shared in the class and which then allow students to discuss and understand each other and to relate the material situations studied to these notions. This type of "milieu" fosters students' construction of ideas focused on the core of the activity, and not its peripheral aspects, as can be observed in some classrooms.

This type of moment in a class is crucial for IBST; it is the core of scientific inquiry. This questioning component supposes that the questions are not only about events (when studying energy, questions such as "will this propeller move?" can be raised) but also about theoretical hypotheses involving a model and concepts ("how much energy is needed to move the propeller?"). If the model elaborated is not relevant, the teacher can then design activities to support students in constructing questions about new science knowledge, relating objects/events and notions/concepts. This epistemic uncertainty can give the opportunity to think at an epistemological level: What are we doing? What types of knowledge are involved – evidence, hypotheses, concepts and laws? It can also provide opportunities to think about the value and degree of certainty that science brings to societal problems. All this thinking can be done because students know enough science to construct new ideas in the framework of an adequate didactic contract and milieu.

These moments are selected from what was going on in the observed classroom. Their analysis aims to propose hypotheses on the conditions of developing scientific inquiry in physics education. Therefore, it is necessary to situate these moments in the type of teaching situations like teacher's introduction of a task, students' working in small groups to carry out a task, or a pooling of the work in small groups, managed by the teacher, etc. The observed moments of "epistemic uncertainty" are situated in two types of teaching situations: when the students are working in small groups and also when the teacher manages a pooling of the work in small groups.

In terms of teacher's actions, this example and the associated comments show that they occur at different points in time: planning the teaching sequence, redesigning activities in accordance with students' actions and understandings, managing the teaching session and reacting on the spot to students' questions or actions to help them think about situations and to be responsible for constructing new ideas involving science knowledge. All these actions necessitate a deep analysis of the knowledge involved in these activities, not only the scientific knowledge but also the knowledge held by students, and a clear overall view of the intended learning outcomes.

3.3 Progression of Knowledge in the Classroom: Institutionalizing the Main Elements Involved in the Previous Class Activity

In Fig. 1, episode 6 that takes place at the whole class level just after the small group work, the teacher initiates a classroom debate. The first stage of this debate is a discussion initiated by the teacher, who describes the diagram proposed by a student on the blackboard and asks the students to give their point of view (this diagram is similar to the diagram presented in Appendix, Fig. 2, part 1, but there is a dotted arrow between the table and the Earth). In the second stage, where two points of view emerged on the actions of the ground and the Earth on an object, such as a book on a table, the teacher intervenes to introduce a scientific point of view; at this moment, she/he takes responsibility for this knowledge.

The teacher institutionalizes the difference between the actions of the ground and the Earth by giving the direction of each action, using verbs like "attracting" and "falling down" in the case of the Earth and "preventing the table from sinking down" for the ground. This institutionalization is a bridge between the knowledge that has been already institutionalized, the ideas developed by the students during the work in small groups and the new elements of knowledge which are currently institutionalized. It should help students to relate these new elements to other elements already acquired. To do that, the teacher uses rational arguments, based on experimental facts that are easily understandable by the students.

In discussions about IBST, institutionalization is rarely mentioned. This is not surprising because IBST is often perceived to be about the nature of science and methods of learning science wherein particular students should be engaged in hands-on activity, but not about classroom management during an academic year. Moreover, institutionalization may be perceived as transmission teaching. In our perspective, these moments of institutionalization, however, regulate the progression of knowledge in the classroom and also introduce knowledge legitimate by the scientific community. From both teacher and student perspectives, the institutionalized elements of knowledge are established, and rather than being considered as questionable, they are themselves used to bring new elements of knowledge into question. Of course, in some cases, these institutionalized elements can also be further questioned, but not in the same way as before, since new questions are fed by the previous elements of knowledge. In a classroom, this progression of knowledge is necessary for effective learning.

When institutionalizing knowledge, the teacher is in the position of a representative of the scientific community; statements are not made from his/her own authority but from the authority of the scientific community. For example, the teacher can say "scientists say that…", or she/he can refer to scientific documents, etc. In such a position, the teacher can argue for these new elements of knowledge, whatever their actual scientific status. In the example given above (episode 6), the argument comes from experimental facts, but in other cases, it might be from a principle based on consensus within the scientific community or from a hypothesis that is still questioned by scientists. Institutionalization is a teaching moment that, depending on the way the teacher proceeds, can give students insight into the ways of the scientific community. Alternatively, it can be reduced to a personal act of authority, if the knowledge is presented as coming from the teacher as a person and not as a representative of the scientific community.

3.4   Students Learning: Differentiating Teaching and Learning Time

The last three episodes, and in particular the last two in session 6, show that some students do not correctly use the knowledge that the teacher institutionalized in session 2. These students are able to use appropriate rational arguments when the teacher invites them to do so, but cannot systematically do this by themselves. Consequently, the teacher, after the moment of institutionalization, manages the students' difficulties, firstly, at the classroom level immediately after the moment (episode 8) and, then individually, when students are working in small groups. The teacher takes the time to help students use the arguments already introduced (episode 9) and also explains them further at the whole class level, but in terms of forces introduced after the institutionalization (episode 10).

More generally, the teacher should be aware of the possible gap between what is taught and what is learnt. In other terms, the institutionalization of an element of knowledge does not imply that students have learnt it. In the classroom, it will be regarded as an established element of scientific knowledge, but it is understood that some students need more time to learn it. In the didactic contract perspective, it also means that new knowledge can be constructed from this previously institutionalized knowledge. Thus, the teacher's management and balancing of teaching time and learning time are not easy. Recognizing this difference allows teachers to use the collective class memory and to adapt their teaching to the students according to their understanding.

In IBST, this difference between teaching time and learning time is rarely discussed. However it is necessary to take it into account if the teacher asks the whole class to propose and discuss new ideas, hypotheses or results, in order that students can understand each other.

Globally, these four conditions facilitate student responsibility for the progression of knowledge in the classroom and the development of students' intellectual autonomy, as we stressed in the discussion of the episodes.

The implementation of these conditions in classroom necessitates some teacher's actions to plan and to teach in the classroom. In the following we specify some of these actions.

4 Teacher's Actions Associated to the Main Teaching Conditions for Students' Intellectual Autonomy

In Table 1 we propose teacher's actions associated to the four teaching conditions presented above. These actions are based on research studies focused on the design of teaching sequences (Tiberghien et al., 2009) and on analyses of classroom practices (Tiberghien & Venturini, 2015; Tiberghien & Venturini, under press). The planning actions aim mainly to design the milieu whereas the classroom actions set up a specific contract with the management of the milieu.

The list of actions is not exhaustive; we present those particularly relevant. They aim at developing students' ownership of knowledge, and thus they develop a continuity of knowledge and a coherent didactic contract. For the first three conditions, the proposed classroom actions correspond to the management of a specific classroom moment situated mainly during small group work, pooling and institutionalization situations; they are at the scale of the duration of this moment that is about some minutes or dozen of minutes. These three conditions (first three lines of Table 1) are sequential even if the teacher's actions associated to a moment may happen incidentally during another one. On the other hand, the last condition leads the teacher to actions which can be done at almost any classroom moment like teacher-student interaction in small group or even a recall during a whole class moment like the two first ones in Table 1. This condition of differentiating teaching and learning time does not correspond to a specific classroom moment. This is why we separate this last condition from the others by a thicker line.

These classroom actions associated to planning actions and the conditions can be studied in teacher's professional development with relevant associated videos (Alonzo, Kobarg, & Seidel, 2012; Cross, 2010; Tiberghien, 2015).

Although our examples concern mechanics, a "traditional scientific theme", these same conditions are also relevant for other types of knowledge such as socio-scientific issues. They are not specific to content, even if they necessitate a deep analysis of it, and favour classroom practices beyond the teaching time of a specific theme.

Table 1 Teacher's action during planning teaching and classroom teaching associated to the main teaching conditions for intellectual autonomy. The thicker line means that the last condition does not correspond to a specific teaching moment

Teaching conditions	Teacher's action	
	Teaching planning	*Classroom teaching*
Sharing some common knowledge and meaningful vocabulary	Choosing the necessary elements of knowledge including the associated representations (e.g. to act, action, diagram system-interaction)	When students work in small group, helping them to raise awareness of the essential elements of knowledge (e.g. to act, action, objects) and helping students to express their ideas
	Designing classroom activities that involve these elements	In whole class, ensuring that students having different propositions intervene and favouring a discussion (the next step is the institutionalization)
Managing moments where an epistemic uncertainty can emerge	Designing classroom activities where main elements of knowledge (according to the content analysis) can be put in question	When students work in small group, helping them to clarify their propositions and to debate them
		In whole class making public the work of some students with different propositions and putting it in debate to bring out rational arguments that could be accepted or rejected (the next step is the institutionalization)
Managing moments of institutionalization	Planning a text and drawings that present the new elements of knowledge	In whole class, proposing the text and drawings of the new knowledge elements to the students whilst relating them to elements already used in the classroom including the developed arguments
Differentiating teaching time and learning time	Planning classroom activities where the elements of knowledge already introduced should be reused	When students work in small group or in whole class:
		Recognizing that students still have not understood elements of knowledge already taught
		Using similar arguments to those already used in the classroom
		Helping students to relate these elements to other elements already acquired

5 Conclusion

In this chapter we presented some main conditions so that the teacher can help students to develop the intellectual autonomy that is a central component of IBST as presented in the introductory section. We analyse and propose some components of teaching practices to mainly develop some cognition and metacognition aspects, which address some constraints relative to teacher professional development.

It showed teacher's actions outside the classroom like planning the academic year, the lesson sequence and the lessons themselves and inside the classroom like managing the course of the session, debates, answering students on the spot and institutionalizing knowledge. These actions should be coherent, in order to develop a didactic contract where students know that teacher expects them to develop new ideas with arguments based on their prior (or previously taught) knowledge and that these ideas should be respected and discussed in the whole classroom. This type of contract needs particular moments in the classroom, and we discussed two of them: moment of "epistemic uncertainty" and institutionalization. Whereas the former allows the presentation and discussion of new ideas that can help to solve problems, the latter allows the teacher to make statements about elements of knowledge on the authority of the scientific community.

The DVD that we have designed provides opportunities to construct and discuss these actions, based on the series of annotated episodes reflecting the dynamics of a class.

Appendix: Extract of the Transcription of a Small Group Working on an Activity in Episode 4

Question of the Activity to Which the Students Answer in Episode 4

Draw the diagram system-interactions of a table on which an object (like a book) is set.

Transcription Extract

(M and C are working together in small group)
… (0:41:15.8)
… (*M and C writes their answer*)

1.M (*stopping writing*) ah but between the Earth and the ground, may be it is not the same because it is on the…it is the Earth it acts the Earth it acts, but it is the ground…that acts do you understand what I mean?
2.C yes
3.M but here
4.C but the ground it is normal, we have the Earth…
5.M …not necessarily…look, imagine that you are on something hard there…it does not act directly on the Earth, if the Earth…
6.C I agree with you but do not go too far; it is like the story of the support…
7.M yeah you're right…
8.C you put the Earth…it is largely enough for [*question above*] b (0:43:33)

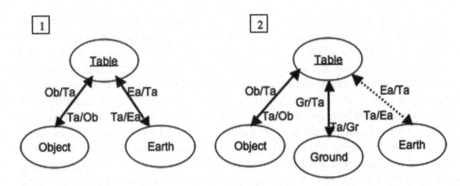

Fig. 2 Part 1, student's solution; part 2, correct solution. In this diagram system-interactions, an ellipse represents an object, a full double arrow a contact action and a dotted double arrow a distant action; the object under study (on which the actions are represented) is underlined (*Ob* means object, *ta* table, *Gr* ground, *Ea* Earth)

References

Alonzo, A. C., Kobarg, M., & Seidel, T. (2012). Pedagogical content knowledge as reflected in teacher–student interactions: Analysis of two video cases. *Journal of Research in Science Teaching, 49*(10), 1211–1239.

Badreddine, Z., & Buty, C. (2011). Discursive reconstruction of the scientific story in a teaching sequence. *International Journal of Science Education, 33*(6), 773–795.

Brousseau, G. (1997). *Theory of didactical situations in mathematics.* Dordrecht, The Netherlands: Kluwer Academic Publishers.

Cross, D. (2010). Action conjointe et connaissances professionnelles. *Éducation & Didactique, 3*(4), 39–60.

Malkoun, L. (2007). *De la caractérisation des pratiques de classes de physique à leur relation aux performances des élèves: étude de cas en France et au Liban.* Doctorat, Université Lyon 2/ Université libanaise, Lyon/Beyrouth.

Mercer, N. (2008). The seeds of time: Why classroom dialogue needs a temporal analysis. *The Journal of the Learning Sciences, 17*(1), 33–59.

Roth, K. J., Garnier, H. E., Chen, C., Lemmens, M., Schwille, K., & Wickler, N. I. Z. (2011). Videobased lesson analysis: Effective science PD for teacher and student learning. *Journal of Research in Science Teaching, 48*(2), 117–148.

Sensevy, G. (2007). Des catégories pour décrire et comprendre l'action didactique. In G. Sensevy & A. Mercier (Eds.), Agir ensemble: Eléments de théorisation de l'action conjointe du professeur et des élèves (pp. 13–49). Rennes, France: Presses Universitaires de Rennes (PUR).

Tiberghien, A. (2015). L'usage de la vidéo en formation: une approche didactique. In L. Ria (Ed.), *Former les enseignants au XXIE Siècle Volume 1. Établissement formateur et vidéoformation.* Louvain-la-Neuve, Belgium: de Boeck.

Tiberghien, A., Cross, D., & Sensevy, G. (2014). The evolution of classroom physics knowledge. *Journal of Research in Science Teaching, 51*(7), 930–961.

Tiberghien, A., & Malkoun, L. (2010). Analysis of classroom practices from the knowledge point of view: How to characterize them and relate them to students' performances. *Revista Brasileira de Pesquisa em Educação em Ciências, 10*(1), 1–32.

Tiberghien, A., & Venturini, P. (2015). Articulation des niveaux microscopiques et mésoscopiques dans les analyses de pratiques de classe à partir de vidéos. *RDST. Recherches en didactique des sciences et des technologies, 11*, 53–78. https://doi.org/10.4000/rdst.986

Tiberghien, A., Vince, J., & Gaidioz, P. (2009). Design-based research: Case of a teaching sequence on mechanics. *International Journal of Science Education, 31*(17), 2275–2314.

Tiberghien, A., & Venturini, P. (under press). Characterisation of the didactic contract using the video of the classroom as primary data. In L. Xu, D. Clark, & G. Aranda (Eds.), *Video-based research in education: Cross-disciplinary perspectives*. Abingdon: Routledge/Taylor & Francis Group.

Valentin, L. (1983). *L'univers mécanique*. Paris: Hermann.

Concluding Remarks: Theoretical Underpinnings in Implementing Inquiry-Based Science Teaching/Learning

Loucas Louca, Thea Skoulia, Olia E. Tsivitanidou,
and Costas P. Constantinou

1 Introduction

The book is a collection of edited chapters on inquiry-based science teaching/learning (IBST/L) with a twofold purpose: to provide resources on the implementation of IBST/L on the one hand and, on the other hand, to highlight ways that those approaches could be promoted in various contexts across Europe, through initial teacher preparation, teacher induction programmes, and professional development activities. To address these points, the book is a compilation of case studies providing a broad range of educational approaches that adopted and made productive use of IBST/L in various countries and educational systems across Europe. Through a variety of approaches reflected in these case studies, the reader can identify the efforts of collaborative groups of science education researchers and practicing science teachers in two areas: (1) applying theoretical ideas in practice and (2) bridging the gap between broad policy perspectives, the specific educational realities of local school traditions, and embedded practices ingrained in the national educational culture.

Before continuing, we would like to highlight an important characteristic of this book. Although the book is about IBST/L, the chapters throughout the book appear somehow isolated from each other. However, we contend that this is one of the strengths of the book. The chapters in this book came together based on the application of the same theoretical framework of IBST/L in various contexts. All authors used and applied the ideas from the IBST/L framework in their own contexts, in

L. Louca (✉) · T. Skoulia
Department of Education Sciences, European University Cyprus, Nicosia, Cyprus
e-mail: L.Louca@euc.ac.cy; T.Skoulia@euc.ac.cy

O. E. Tsivitanidou · C. P. Constantinou
Department of Educational Sciences, University of Cyprus, Nicosia, Cyprus
e-mail: tsivitanidou.olia@ucy.ac.cy; c.p.constantinou@ucy.ac.cy

© Springer International Publishing AG, part of Springer Nature 2018
O. E. Tsivitanidou et al. (eds.), *Professional Development for Inquiry-Based Science Teaching and Learning*, Contributions from Science Education Research 5, https://doi.org/10.1007/978-3-319-91406-0_14

their school settings, and in their countries. Thus, the various chapters presented a pluralistic picture of application of the same set of theoretical ideas about IBST/L in a number of different contexts and therefore provide different results. On the other hand, the chapters are homogeneous about inquiry and IBST/L.

For this reason, in the introductory chapter, the authors summarized the various characteristics of inquiry in science education, provided an overview of the educational policy priorities formulated at the European level for IBST/L, and discussed the opportunities and constraints that these efforts have generated for science education and science teacher professional development across European contexts. Based on that, the various chapters as a whole comprise good examples of the outcomes of the application of the idea of IBST/L to various educational systems and educational levels throughout Europe. In this view, the reader can read this collection of chapters as a collection of case studies of various different efforts of applying IBST/L throughout Europe.

Therefore, our purpose in this concluding chapter is to provide the reader with a structured overview of the main ideas described and discussed throughout the book, focusing on four main topics that underpin the chapters in the book. First, using examples described in various chapters of the book, we provide an overview of what scientific inquiry should look like in authentic learning environments. This provides the reader with a review of the various facets of inquiry presented throughout the book. Second, we describe several theoretical frameworks from the literature underpinning IBST/L as used in the chapters throughout the book, seeking to give the reader a better overview of the theories related to IBST/L implementations. Third, we provide an overview of pedagogical content knowledge (PCK) as one of the many frameworks that can unite efforts of professional development in IBST/L as presented in this book. This aims to help the reader situate all the efforts described in the various chapters of the book within the same theoretical framework of what inquiry-based approaches in science education require from teachers in terms of knowledge and abilities. Lastly, in Sect. 5, we highlight effective strategies for professional development to help teachers implement IBST/L.

2 What does Inquiry Look Like in the Science Classroom?

The literature from a variety of perspectives and intellectual traditions concerning children's abilities in scientific inquiry shows a general consensus regarding the things we should value and promote in children's inquiry (e.g. Linn, Davis, & Bell, 2004; Minstrell & Zee, 2000; NRC, 2007; Osborne, Erduran, & Simon, 2004). However, this consensus does not extend to defining what scientific inquiry looks like in the science classroom. Rather, it contends against a widely shared sense of inquiry as a pedagogical strategy, that is, a method for teaching the traditional 'content' of science (e.g. Hammer, 1995, 2004). According to this view, assessing the quality of children's inquiry is equivalent to assessing their progress towards the correct answers.

Even in cases where inquiry is valued as a process of developing scientific thinking, there is still no consensus regarding what inquiry exactly entails. For many, inquiry is an effective method for learning science content, whereas others emphasize it as a 'part of science' and, thus, as a teaching objective in itself (Louca & Zacharia, 2007). When inquiry is the method for learning, it is at best a valuable teaching tool more productive than traditional approaches of knowledge. When inquiry is part of science education, then teaching includes helping children to understand its nature and develop abilities to use scientific inquiry effectively for learning, in addition to learning about scientific phenomena themselves. By inquiry, we refer to the 'activities of students in which they develop knowledge and understanding of scientific ideas', as well as an understanding of how to study the natural world (NRC, 1996, p. 23). To offer a more specific definition, we take inquiry to mean the pursuit of causal, coherent explanations of natural phenomena (Hammer, 2004). This may include a variety of classroom-based forms of inquiry, both activity-based (e.g. designing experiments and controlling variables, collecting and interpreting data and observations from physical phenomena) and discourse-based (e.g. using data or observations to engage in argumentation and to engage in mechanistic and analogical reasoning).

The definition of children's inquiry that we adopt for this chapter suggests the relationship between inquiry as an activity and inquiry as discourse. In fact, from a sociolinguistic perspective (Carlsen, 1991), educational research has stressed the instructional functions of classroom-based inquiry in science as a means for facilitating the construction of scientific knowledge (Solomon, 1994) and supporting children's abilities in scientific reasoning (e.g. May, Hammer, & Roy, 2006; Russ, Scherr, Hammer, & Mikeska, 2008). Following this, a growing body of research has developed an interest in classroom discourse (e.g. Abell, Anderson, & Chezem, 2000; Cazden, 2001; Edwards & Westgate, 1994; Hogan, 1999; Kelly & Crawford, 1997; Lemke, 1990) for its relevance to children's inquiry (Hammer, 1995; van Zee, 2000; van Zee & Minstrell, 1997), for the development of student ideas in science (Mortimer, 1995), and for students' conceptual and cognitive development (Sprod, 1998). It is important to differentiate between children's inquiry (both activity- and discourse-based) and non-focused exploratory talk. We define discourse-based inquiry to include not only knowledge claims and ideas but also children's reasoning and inquiry processes (Chin, 2006), such as children's questions and comments and children's epistemologies or experiences used to support their ideas or thinking (Louca, Tzialli, & Zacharia, 2012).

Over the years, descriptions of good examples for inquiry-based learning in science education have been put forward by researchers, teacher educators, and experienced teachers. However, attempts to change the dominant deductive teaching style have proved to be highly challenging. Teachers' professional competences and teaching approaches are of crucial importance for keeping a proper balance between instruction and autonomous construction in the teaching and learning of science. The literature in this area suggests that science teachers' instructional practices have a crucial influence in preserving this balance. It also suggests that professional development is needed to help in maintaining this balance.

In addition, due to the fact that IBST/L approaches are, in many cases, at the core of teachers' motivational thinking, science teachers tend to assume that inquiry approaches automatically benefit students' motivation. Much of this could be possibly be ascribed to a general lack of knowledge regarding motivational theory among teachers. To unfold motivational potential in their students, teachers must have the capacities to design and enact a range of IBST/L approaches. More specifically, they should be able to include motivational strategies in their planning of IBST/L, be aware of individual students' motivational states, and be capable of responding to motivational issues that arise in classroom situations. Consequently, much could be achieved if teachers acquired theoretically informed knowledge about motivation and strategies for incorporating this in their planning and implementation of IBST/L.

Towards this direction, Bungum in chapter "Science Inquiry as Part of Technological Design: A Case of School-Based Development in Norway" described a project developed and run by teachers in a lower secondary school in Norway, where students in grade 9 spent 2 weeks designing and building individual car models. The application of science and other subjects in a creative technological context has great potential for familiarizing young people with the overlap of science and technology. Innovations in this direction require fundamental changes in how we look upon the aims and content of school science. Authentic inquiry can be achieved by allowing for students' creativity, setting high expectations for product quality, and providing enough time for students to participate in creative inquiry processes. To experience what it means to work with science in a modern society, students should be given opportunities to engage in science and inquiry in purposeful, creative activities and to develop their skills related to a technological outcome. Thus, in her chapter, Bungum discussed ideas about how inquiry-based learning can provide students with experiences of purposeful inquiry forms as part of the development of high-quality technological products. Based on this case, Bungum argued that inquiry-based teaching calls for a certain level of autonomy for learners. The success of the project lies in its representation as an exception to normal curriculum structure and reallocation of resources to achieve goals set by the school. These goals transcend the narrower goals of the formal curriculum and contribute to combating current trends of increased rigidity and narrow focus on testable knowledge in schools.

Chapter "Promoting IBSE Using Living Organisms: Studying Snails in the Secondary Science Classroom" provided a paradigmatic example of IBST/L in the context of biology and environmental education. In this chapter, Rybska suggested ways to promote IBST/L in secondary school level of biology education, through the interaction of students with snails. In this case study, Rybska used snails to explore the extent to which IBST/L may be promoted through hosting living organisms in the classroom. Even though children generally appreciate animals, snails commonly invoke feelings of disgust. Rybska described a rigorous development and implementation of a teaching-learning sequence (TLS) with emphasis on reflection on students' own learning to guide teaching. The resulting TLS followed a process of observing, hypothesizing, collecting and analysing evidence, and engaging in a debate about the interpretation of the data. Implementation findings revealed an improvement in students' attitudes and some changes in students' knowledge, especially on ecological aspects.

In chapter "Drama As a Learning Medium in Science Education", Peleg and colleagues proposed drama in science learning as an alternative teaching method for triggering students' motivation towards science. The authors suggest that drama within science education proposes an innovative way of introducing IBSL in a science class, which can be also used for triggering students' interest and motivation towards science learning, addressing the need for a more motivational school science instruction, which might stimulate students' interest in science and science careers (e.g. Osborne & Dillon, 2008). Following a recent call for 'innovative curricula and ways of organising the teaching of science that address the issue of low student motivation' in Europe (Osborne & Dillon, 2008, p. 16), chapter "Drama As a Learning Medium in Science Education" provided science teacher educators with theoretical and practical knowledge of how drama can serve as an inquiry-based teaching and learning tool in science education, to increase students' scientific literacy, engagement, and motivation.

The authors outlined two important and necessary elements that need to be considered for successful teaching discourse (Dorion, 2009). The first involved the simulation of social events (O'Toole & Dunn, 2002), which allows for exploration of how people behave in other human contexts. Activities whereby the learners take on a human character were referred to as 'simulations' (McSharry & Jones, 2000). The second strategy employed miming and role play for presenting abstract physical phenomena, which were referred to as 'analogies' (McSharry & Jones, 2000). In such cases, students no longer act as humans but rather as physical or natural entities, such as molecules, animals, and photons. The authors also suggest that drama activities in education are characterized by the following aspects: (a) drama uses fiction, (b) it allows for mental models to be constructed and examined, (c) drama activities are sociocultural activities and allow for scaffolding in learning science, (d) these activities introduce imagination and creativity to the science classroom, (e) they allow for both narrative and logical scientific thinking, and (f) they can stress process, performance, or both. All of the above suggest that drama can be an inquiry-based learning method that functions through narrative and is multimodal, multisensory, and sociocultural, providing examples of successful activities of drama in science education.

In chapter "Designing Teacher Education and Professional Development Activities for Science Learning", Tiberghien, Badreddine, and Cross presented resources for teacher development on the basis of classroom practices that favour students' intellectual autonomy. They described in detail a video analysis of a 10th grade physics classroom teaching sequence on mechanics. Their analysis was presented in relation to the students' progression of knowledge, its continuity, and scientific practices developed in the classroom. From that analysis, students' ownership was revealed as an essential aspect of learning in IBST/L, which subsequently leads to a degree of intellectual autonomy. The authors also identified four teaching conditions that aim in developing students' intellectual autonomy, including a responsibility for sharing knowledge, managing moments of 'epistemic uncertainty', institutionalizing the main elements of knowledge involved in the previous class activity, and differentiating teaching and learning time.

The idea of incorporating meaningful contexts, relevant for citizens of a modern society, has already been underlined in science education. IBST/L can offer such meaningful contexts in science education, but its potential can be best realized through constructive collaboration between teachers and across subjects. Several different and innovative perspectives for introducing IBST/L in science classrooms, from cross-curricular projects in science and technology education, are presented and discussed throughout the book. Evidence from empirical research in science classrooms on the potential benefits that IBST/L might bring into learners' cognitive, metacognitive, and emotional domains is presented.

3   Theoretical Frameworks Underpinning IBST/L

Difficulties in defining forms of IBST/L may be addressed by adopting particular theoretical lenses that can help the reader, science teachers, researchers, and teacher educators and trainers to obtain a theoretical perspective of inquiry-based learning, making it easier to identify inquiry in authentic, classroom-based learning environments (e.g. Hammer, 1995, 2004) and, in particular, helping teachers assess the quality of children's inquiry.

The authors in this book adopted a repertoire of theoretical perspectives, in an effort to cover the spectrum of IBST/L more efficiently. Below, we summarize the most dominant theoretical perspectives, with a note that these are not the only ones available in the literature. The theoretical perspectives discussed as an integral part of teaching science as inquiry, or viewing science teaching as inquiry, include theory and research in motivation, self-efficacy, scientific literacy, dialogic teaching, the communicative approach, and the nature of science.

3.1   Motivational Theories

In chapter "Using Motivational Theory to Enrich IBSE Teaching Practices", Andersen and Krogh adopted what they call a 'pragmatic use' of motivational theory, which includes the extraction of motivational constructs relevant to science education and provides motivational foci for teachers' planning of IBST/L practices. They suggested that the realization of motivation theory in science teachers' training ensures that a pragmatic selection of motivational knowledge emphasizes its usefulness for science teachers. The authors provided descriptions of various strategies, which have been proven fruitful in facilitating a transformation of theoretical motivational knowledge into propositionally interpreted practical knowledge and enriched teaching practices. They suggested that to unlock the motivational potential in science education, teachers must have the abilities and resources to design and enact a range of IBST/L approaches that include motivational strategies, be aware of individual students' motivational states, and be capable of responding to motivational issues that arise in classroom situations.

Looking at the motivational mechanisms available, the authors suggested that two different timescales are operating: constructs like autonomy and task value may be turned on and off in relation to singular situations, whereas self-efficacy, relatedness, causal attributions, and goal orientations to some extent can be influenced by situations but have to be built by consistent efforts over longer periods. Relying on the motivational framework of CARTAGO (Competence/Self-efficacy, Autonomy, Relatedness, Task Value, Attributions to success or failure, and Goal Orientations Mastery vs. Performance), Andersen and Krogh identified a number of recommendations for a motivational classroom practice, which serve as a valuable tool for motivational planning and analysis of IBST/L approaches. The authors also provided a description of some major design principles and strategies, such as reflective writing, and awareness activities, which are used to facilitate teachers' transformation of theoretical inputs into classroom practice.

3.2 Interaction of Self-Efficacy and Scientific Literacy

In chapter "Taking Advantage of the Synergy Between Scientific Literacy Goals, Inquiry-Based Methods and Self-Efficacy to Change Science Teaching", Evans and Dolin described the development and implementation of teacher professional development workshops that facilitated the use of IBST/L methods to achieve scientific literacy and active enhancement of self-efficacy. The basic strategy was that by empowering teachers' educators and teachers to work towards more motivating scientific literacy goals using the kind of inquiry described by MTG (2007), teachers' self-efficacy for science teaching could be actively enhanced more readily than by using traditional teaching methods with less motivating content (Andersen, Dragsted, Evans, & Sørensen, 2004).

In this sense, IBST/L teaching environments and approaches interact with and modify teachers' beliefs about their science teaching self-efficacy and, consequently, enhance the quality of science teaching and learning (Andersen, Dragsted, Evans, & Sørensen, 2005). Since self-efficacy could be related to successful science teaching (e.g. Enochs & Riggs, 1990; Woolfolk & Hoy, 1990), changes in self-efficacy may be useful in helping science teachers become more successful. Over the years, research has found evidence for the relationship between self-efficacy and effective teaching (e.g. Ashton & Webb, 1986; Czerniak, 1990; Enochs & Riggs, 1990; Gibson & Dembo, 1984; Guskey, 1988; Woolfolk & Hoy, 1990). It has also shown that teacher self-efficacy beliefs strongly influence the nature of a teacher's role, planning, and, consequently, curriculum and student learning (Tobin, Tippins, & Gallard, 1994).

In this vein, Ford (1992), for instance, suggested that patterns of personal agency beliefs, such as personal capability and context, may determine behaviour. He proposed a taxonomy of several patterns to represent the interactions of personal capability and context and hence to understand the likely behaviour of individuals with various belief combinations. Using sophisticated inquiry methods to teach scientific literacy objectives in a climate that promotes self-efficacy may provide a combination of context and beliefs which is motivating for teachers. It is the intrinsic motivation of a higher self-efficacy for using IBST/L which increases the likelihood of teachers both attempting and continuing to use IBST/L methods.

In 1997, Bandura suggested detailed mechanisms by which the self-efficacy of teachers can be maintained, raised, or diminished. He strengthened the link between self-efficacy and the extent to which it is influenced by the context of the situations in which teaching is performed. He also differentiated self-efficacy from other, less malleable constructs, such as self-confidence and self-concept, which are both more general and less situation-specific. Bandura proposed a number of mechanisms by which teacher self-efficacy may be influenced: (1) mastery of teaching experiences, (2) modelling by other teachers, (3) authentic and valid performance feedback, and (4) environments where stress is not inhibiting. Consequently, Evans and Dolin's teacher professional development package consciously included all of these strategies.

3.3 Dialogic Teaching and the Communicative Approach

Learner participation, motivation, and deeper learning are of high importance in a dialogic approach for teaching (Alexander, 2006), in the context of which a fundamental aim of dialogic teaching is to explicitly extend student reasoning and understanding. The authors of chapter "Inquiry-Based Approaches in Primary Science Teacher Education" (Lehesvuori, Ratinen, Moate, & Viiri) claimed that the dialogic approach does not adequately stress the authoritative aspect of science education. They proposed the communicative approach as an essential dimension of IBST, which can potentially address the gap between students' pre-existing views/ideas and scientifically accepted knowledge. In this framework, the concepts of dialogic teaching and the communicative approach in the context of inquiry-based learning (Mortimer & Scott, 2003) are embraced. The authors of chapter "A Teacher Professional Development Programme on Dialogic Inquiry" (Enghag, Engström, & Brorsson) also claimed that the development of a dialogic science classroom is a way to enhance teaching and learning, where students of different backgrounds, languages, and experiences create the classroom environment (e.g. Martin, 1993; Mortimer & Scott, 2003; Wells, 1999).

The main principles of IBST/L approaches are related to the fundamental ideas of dialogic teaching. For instance, including dialogic and communicative approaches in science education addresses concerns about the lack of openness in inquiry-based approaches, i.e. learners working towards predetermined outcomes (Sadeh & Zion,

2009). Authenticity and openness cannot be conveyed through applying transmissive and authoritative forms of interaction.

A fundamental aim of dialogic teaching is to explicitly extend student reasoning and understanding, which are essential in dialogic teaching. The five key characteristics of the dialogic approach (Alexander, 2006) can be briefly described as follows:

(a) Collective: teacher and students jointly participate in the learning as a group or as a class.
(b) Reciprocal: teacher and students listen to each other, share ideas, and consider alternative views.
(c) Supportive: students can present their ideas freely without fear of being incorrect.
(d) Cumulative: teacher and students develop their ideas together, jointly constructing knowledge.
(e) Purposeful: the teacher plans and guides the discourse, paying attention to educational goals in addition to the above points.

As Enghag, Engström, and Brorsson indicated, an open classroom context, where feedback from teachers and other students is essential, is fundamental to dialogic inquiry. If someone actively responds to a statement, an answer, or a question posed, this will provide understanding to a much greater extent than when one-way communication is the predominant form of discourse (Bakhtin, 1981, 1986).

To create a classroom where dialogue is essential, it is important that the teachers are aware of how to use different communicative approaches (Mortimer & Scott, 2003). Mortimer and Scott's communicative framework accommodates both dialogic and authoritative approaches in the science classroom. According to Mortimer and Scott (2003), classroom discourse consists of four categories generated from the combination of two dimensions: interactive/noninteractive and authoritative/dialogic (Mortimer & Scott, 2003). Using these categories, the communicative approach addresses both the everyday understanding or prior knowledge of learners and the authoritative view of science. The interactive/noninteractive dimensions indicate the different ways in which teachers can use talk, whether through whole-class discussions, question/answer sessions, or teacher talk. Scott and Ametller (2007) stressed that meaningful science teaching should include both dialogic and authoritative aspects and that the relationship between these two aspects is highly significant.

3.4   The Nature of Science as an Integral Part of Teaching Science as Inquiry

In their chapter, Cakmakci and Yalaki argued that IBST/L is closely aligned with the nature of science (NOS), and thus it should not be separated from teaching and learning about NOS in school science. In this sense, the authors suggested that teacher professional development can help teachers to be trained to utilize IBST/L to support the development of students' understanding of NOS. The authors also suggested that understanding NOS is an essential aspect of public engagement with science and scientific literacy (Driver, Leach, Millar, & Scott, 1996); even though IBST/L has been shown to be an effective way of teaching science (Minner, Levy, & Century, 2010), it does not necessarily help students to understand NOS or achieve scientific literacy. The NOS aspects of IBST/L are consistent with views emphasized in recent policy and reform documents in science education (e.g. NSTA, 2000). Research suggests that students who engage in scientific inquiry alone do not necessarily develop a contemporary understanding of NOS (Lederman, 1999). Therefore, researchers have usually used an explicit and reflective approach for developing students' NOS views rather than an implicit approach that utilizes hands-on or inquiry science activities lacking explicit references to NOS (Lederman, 2007). In this respect, teaching about NOS should be addressed explicitly and reflectively within contextualized activities rather than only within generic (decontextualized, domain-general) activities (Cakmakci, 2012; Clough, 2006; Duschl, 2000; Leach, Hind, & Ryder, 2003).

4 Developing Teacher Competencies in IBST/L

Chapters throughout the book described case studies in science education and IBST/L based on particular contexts, including teacher professional development. In this section, we provide a detailed review of the literature on pedagogical content knowledge (PCK), proposing the use of PCK as a productive framework that can unite efforts of teachers' professional development in IBST/L as presented in this book. Authors in this book have used (explicitly or implicitly) – but not all in a single chapter – a variety of the characteristics of PCK that we present below, supporting the notion that PCK is a rigorous theoretical framework that can provide a basis for the development of teacher professional development programmes.

PCK is a conceptual framework which was developed to help understand the content and nature of teacher knowledge (Alonzo & Kim, 2016). PCK is considered as the professional knowledge base of teachers – for example, science teachers – the possession of which differentiates teachers from scientists (Cochran, DeRuiter, & King, 1993; Veal & MaKinster, 1999). In this view, PCK is defined as the 'teacher capacity' to transform the content knowledge 'into forms that are pedagogically powerful and [...] adaptive to the variations in ability and background' of the students (Shulman, 1987, p. 15). It includes knowledge of how specific subject matter and problems can be organized, represented, and adapted to different interests and abilities of students and used in teaching practice (Magnusson, Krajcik, & Borko, 1999).

This book's purpose was to provide resources on the implementation of IBST/L as well as to highlight ways that those approaches could be promoted in various contexts across Europe, through initial teacher preparation, teacher induction programmes, and professional development activities. To succeed in any teacher education programmes, teacher educators need to address particular teacher needs in productive ways. PCK may provide a theoretical framework that teacher education readers of this book may use to read the case studies described in the book chapters, relating descriptions to teachers' pedagogical knowledge (or needs) to their subject matter knowledge and the school context.

The book is a compilation of case studies providing a broad range of educational approaches that adopted and made productive use of IBST/L in various countries and educational systems across Europe. Due to that, we suggest that by identifying areas of teacher expertise (or learning needs for that matter) that are related to the various implementations of IBST/L, the reader may find ways to adopt ideas described throughout the book to different teacher contexts.

In this particular sense, 'PCK is […] a tool for describing and contributing to our understanding of teachers' professional practices' (Kind, 2009, p. 198). It refers to the knowledge that teachers develop on how to teach a specific content in particular ways in order to lead to enhanced student understanding (Loughran, Berry, & Mulhall, 2006).

Since its introduction back in 1986 by Shulman, the concept of PCK has been adopted by various researchers and has led to both theoretical developments and empirical research (Evens, Elen, & Depaepe, 2016). As a conceptual construct, PCK has been interpreted in different ways by different researchers resulting in different PCK models over the years (Cochran et al., 1993; Fernández-Balboa & Stiehl, 1995; Grossman, 1990; Hashweh, 2005; Kind, 2009; Koballa, Gräber, Coleman, & Kemp, 1999; Magnusson et al., 1999; Marks, 1990). Therefore, there is no common conceptualization of PCK (Abell, 2007; Smith, 1999; Van Driel, Verloop, & de Vos, 1998).

Interpretations mainly differ in the definitions of the individual components of PCK, in how these components are linked to each other, and in the relationship between subject matter knowledge (SMK) and PCK. Shulman's view is that SMK is a separate knowledge base. Some researchers (e.g. Grossman, 1990; Magnusson et al., 1999) follow Shulman's line of thought, identifying PCK as a special kind of knowledge used by teachers to transform their SMK to benefit students (Kind, 2009), whereas others (e.g. Fernández-Balboa & Stiehl, 1995; Marks, 1990) include SMK in their definition of PCK.

Shulman's (1986) definition refers to two basic elements: (1) knowledge of representations of the specific content and instructional strategies and (2) understanding of learning difficulties and students' conceptions of specific content knowledge. The first element refers to strategies that teachers use to make subject matter understandable to students (e.g. illustrations, analogies, explanations, and demonstrations), whereas the second refers to teachers' understanding of students' misconceptions, students' ideas gained through the interpretation of previous learning experiences or preconceived ideas about a topic, as well as knowledge of any other barriers related to the learning of the subject matter (Shulman, 1986; Kind, 2009). Therefore, in order to choose teaching strategies to support their student's learning, teachers need to identify and respond to the needs of their students by understanding their learning difficulties and perceptions and by selecting different ways of representing the subject matter (Alonzo & Kim, 2016). Despite the fact that various researchers have expanded Shulman's PCK definition by adding new components, they all agree with the two previously mentioned components of PCK (van Driel et al., 1998).

Grossman (1990) argues that a teacher's knowledge base includes (1) pedagogical knowledge, (2) subject matter knowledge, (3) pedagogical content knowledge (PCK), and (4) knowledge of context. PCK is seen as the transformation of pedagogical knowledge, knowledge of context, and specific content (Fernandez, 2014). In defining PCK, Grossman adds two more components: (1) conception of purposes for teaching subject matter and (2) curricular knowledge. This model is one of the most cited models in PCK studies (e.g. Akkoç & Yeşildere, 2010; Magnusson et al., 1999).

Tamir (1988) adds the knowledge of assessment to PCK models; this component consists of knowledge of dimensions of science learning that are important to assess and knowledge of methods for assessing learning (Tamir, 1988). Marks (1990) suggests that PCK consists of four components which are connected to Shulman's model by adding (1) subject matter knowledge and (2) media for the instruction.

Magnusson and co-authors (1999) highlight the dynamic nature of PCK, mentioning that 'development of PCK is not a straightforward matter of having knowledge; it is also an intentional act in which teachers choose to reconstruct their understanding to fit a situation' (p. 111). Building on previous studies (e.g. Grossman, 1990; Tamir, 1988), they present five components of PCK for science teaching which are (1) orientations towards science teaching, (2) knowledge and beliefs about the science curriculum, (3) knowledge and beliefs about students' understanding of specific science topics, (4) knowledge and beliefs about assessment in science, and (5) knowledge and beliefs about instructional strategies for teaching science (p. 97).

Cochran et al. (1993) defined PCK as 'the manner in which teachers relate their pedagogical knowledge to their subject matter knowledge in the school context, for the teaching of specific students' (p. 1). Considering that teachers' knowledge formed through their teaching practice, Cochran et al. proposed the term pedagogical content knowledge (PCK). The dynamic nature of this knowledge includes four components: (1) subject matter knowledge, (2) knowledge of general pedagogy, (3) knowledge of context, and (4) knowledge of students. According to Cochran et al., 'integration of the four components comprises PCK' (p. 268), and these components must be learned and applied simultaneously in all learning experiences, but not necessarily in equal parts (Nilsson & Vikström, 2015).

5 Effective Teacher Professional Development Strategies

Features associated with improved teaching have been identified from a proliferation of research in teacher professional development (TPD) (e.g. Desimone, 2009; Guskey, 2000). Although multiple ways of organizing these features have been proposed, instead of describing the diverse TPD model possibilities, we focus our discussion below on five important features of TPD based on prior research. These include (1) the content of the TPD, (2) the organization and the structure of the TPD activities, (3) the degree of consistency between new experiences provided to teachers and the national standards, (4) the context in which the TPD takes place, and (5) the degree to which the activities in TPDs emphasize the collective participation of teachers. Our discussion considers the strengths and weaknesses of each TPD feature.

The content focus of teacher learning is one of the most influential features of professional development (PD) (Desimone, 2009), including the overall focus of PD and which teacher knowledge, skills, and experiences are targeted. This can range between focusing on the science content, pedagogical content knowledge, and teaching strategies (Carlsen, 1993; Cronin-Jones, 1991; Hollon, Roth, & Anderson, 1991) or a mixture of these (Hill, Rowan, & Ball, 2005; Penuel, Fishman, Yamaguchi, & Gallagher, 2007). Research suggests a relationship between PD activities – focusing on subject matter content as well as how students learn that content – with improvements in teacher knowledge, skills, teaching practice, and student achievement (e.g. Cohen & Hill, 2001; Desimone, Garet, Birman, Porter, & Yoon, 2002; Garet, Porter, Desimone, Birman, & Yoon, 2001).

A second PD feature is the organization and the structure of the PD activities and the ways teachers are engaged in them. Research has found that reform-based PD formats of study groups, teacher networks, mentoring relationships, internships, or teacher research centres (Opfer & Pedder, 2011) are much more likely to lead to sustainable teacher change than traditional learning formats, such as one-off workshops, limited-time courses, and conferences, which function more as 'style shows' (Ball, 1995; Hawley & Valli, 1999). Moreover, reform activities often take place during the regular school day as part of the process of classroom instruction or during regularly scheduled teacher planning time. On the other hand, research has shown that one-time PD workshops, such as a conference and 1 or 2 day courses, often take place outside the school context, are not typically aligned with the ongoing teaching practice, and do not reliably lead to changes in classroom teaching (Loucks-Horsley & Matsumoto, 1999). Workshop formats of PD are generally criticized for being ineffective in providing teachers with sufficient time, activities, and content necessary to increase their knowledge and promote meaningful changes in their teaching practices (Loucks-Horsley, Hewson, Love, & Stiles, 1998). Engaging teachers in active learning is also related to the effectiveness of PD (Garet et al., 2001; Loucks-Horsley et al., 1998). Active learning may take a number of forms, such as observing expert teachers or being observed, providing or receiving interactive feedback, and reflecting upon student work (e.g. Banilower & Shimkus, 2004; Borko, 2004; Carey & Frechtling, 1997; Darling-Hammond, 1997).

A third PD feature is the degree of consistency between new experiences provided and the state or national standards and curriculum. Desimone with co-authors (2002) suggest that productive PD needs to provide teachers with rich and diverse experiences related to the novel ideas on which the PD focuses. The degree of consistency may also be enhanced by incorporating experiences that are consistent with the participating teachers' goals.

A fourth feature is the context of PD, which includes formality, voluntary participation, and duration. In more formal contexts, PD can be made available through external expertise in the form of courses, workshops, or formal qualification programmes or, alternatively, through more informal forms that may include collaboration at school or teacher level, both within and across schools (Gaible & Burns, 2005; OECD, 2009). Another aspect of context is the choice of participation. Participants who volunteer differ from teachers required to participate, in terms of their motivation to learn, their commitment to change, and their willingness to be risk takers (Loughran & Gunstone, 1997; Supovitz & Zeif, 2000), and subsequently, this impacts the results of PD (Yamagata-Lynch, 2003). Additionally, the needs of volunteers and non-volunteers may differ substantially (Lawless & Pellegrino, 2007). When teachers volunteer to participate in PD programmes, the expectations and requirements for work-related activities increase (Yamagata-Lynch, 2003). Finally, research suggests that teachers need time to develop, absorb, discuss, and implement new practice and knowledge (Garet et al., 2001), relating to both the span of time over which the PD activities are spread and the number of hours spent in the PD activities (e.g. Cohen & Hill, 2001; Desimone, 2009; Garet et al., 2001; Opfer & Pedder, 2011). Once teachers begin to apply new knowledge and skills to their practice, short PD programmes usually offer only a limited follow-up (Penuel et al., 2007), fail to meet the ongoing pedagogical needs of teachers, and are rather disconnected from day-to-day teaching practice (Gross, Truesdale, & Bielec, 2001).

A fifth PD feature is the degree of collective participation of teachers in PD programmes. Research suggests that PD is more effective in affecting teacher learning and practice if teachers from the same school, department, area, or student grade-level participate collectively (e.g. Desimone et al., 2002; Garet et al., 2001; Wayne, Yoon, Zhu, Cronen, & Garet, 2008). Change in teaching behaviour then becomes an ongoing and collective responsibility (Cochran-Smith & Lytle, 1999; McLaughlin & Talbert, 1993; Opfer & Pedder, 2011; Thomas, Wineburg, Grossman, Myhre, & Woolworth, 1998) and can be enhanced by extending collaboration between teachers, school-based teacher mentors, university researchers, and curriculum developer mentors (Gerard, Varma, Corliss, & Linn, 2011).

The role of the PD facilitator is also crucial (Borko, 2004). The support for teachers to clarify ideas and reflect on practice depends on the expertise of the mentor or collaborator and the time allocated for teachers to work with him/her during the PD programme (Cleland, Wetzel, Zambo, Buss, & Rillero, 1999; Ketelhut & Schifter, 2011; Penuel & Yarnall, 2005; Penuel, Fishman, Gallagher, Korbak, & Lopez-Prado, 2008; Williams, 2008). Furthermore, facilitators must be able to establish a community of learners in which inquiry is valued, and they must structure the learning experiences for that community (e.g. Phillips, 2003; Remillard & Geist, 2002; Strahan, 2003).

6   Conclusions

In this chapter we provided an overview of the main ideas discussed throughout the book, seeking to help the reader situate all the efforts of IBST/L within a theoretical framework of what inquiry-based approaches in science education look like and what they require from teachers in terms of knowledge and abilities. We focused on four main topics that underpin the chapters in the book:

1. What does scientific inquiry look like in authentic learning environments?
2. The six main theoretical frameworks underpinning IBST/L throughout the book: theory and research in motivation, self-efficacy, scientific literacy, dialogic teaching, the communicative approach, and the nature of science.
3. Presentation of PCK as a productive framework that can unite the efforts of teachers' professional development in IBST/L as presented in this book.
4. Description of the different ways, tools, and strategies for helping teachers implement this approach for teaching science.

All the above are meant to accompany the edited chapters in the book, presenting the efforts of a number of collaborative groups of science education researchers and practicing science teachers to put theoretical ideas into practice, to bridge the gaps between broad policy perspectives and the specific educational realities of local school traditions, as well as embedded practices ingrained in the national educational culture. Our purpose was to situate all these efforts in different science education disciplines and contexts, in a theoretical framework of teacher knowledge (PCK), principles of designing teacher professional development programmes, and theoretical ideas that cut across IBST/L. These ideas, which include motivation, self-efficacy, scientific literacy, dialogic teaching, the communicative approach, and the nature of science, are widely used in the book. Our ultimate aim was to provide a concluding chapter to a book that initially discussed what inquiry is and its characteristics and also to highlight ways of how to support this type of IBST/L in the classroom through TPD programmes.

References

Abell, S. (2007). Research on science teachers' knowledge. In S. K. Abell & N. G. Lederman (Eds.), *Handbook of research on science education* (pp. 1105–1149). Mahwa, NJ: Lawrence Erlbaum.

Abell, S. K., Anderson, G., & Chezem, J. (2000). Science as argument and explanation: Exploring concepts of sound in third grade. In J. Minstrell & E. H. van Zee (Eds.), *Inquiring into inquiry learning and teaching in science* (pp. 65–79). Washington, DC: AAAS.

Akkoç, H., & Yeşildere, S. (2010). Investigating development of pre-service elementary mathematics teachers' pedagogical content knowledge through a school practicum course. *Procedia-Social and Behavioral Sciences, 2*(2), 1410–1415.

Alexander, R. J. (2006). *Towards dialogic teaching: Rethinking classroom talk.* Cambridge UK: Dialogos.

Alonzo, A. C., & Kim, J. (2016). Declarative and dynamic pedagogical content knowledge as elicited through two video-based interview methods. *Journal of Research in Science Teaching, 53*(8), 1259–1286.

Andersen, A., Dragsted, S., Evans, R. & Sørensen, H. (2005, August). *The relationship of capability beliefs and teaching environments of New Danish elementary teachers of science to teaching success.* Paper presented at the European Science Education Research Association Conference, Barcelona, Spain.

Andersen, A. M., Dragsted, S., Evans, R. H., & Sørensen, H. (2004). The relationship between changes in teachers' self-efficacy beliefs and the science teaching environment of Danish first-year elementary teachers. *Journal of Science Teacher Education, 15*(1), 25–38.

Ashton, P. T., & Webb, R. B. (1986). *Making a difference: Teachers' sense of efficacy and student achievement.* New York: Longman.

Bakhtin, M. (1981). Discourse in the novel. In C. Emerson & M. Holquist (Eds.), *The dialogic imagination. Four essays.* Austin, TX: University of Texas Press.

Bakhtin, M. (1986). The problem of speech genres. In C. Emerson & M. Holquist (Eds.), *Speech Genres and Other Late Essays.* Austin, TX: University of Texas Press.

Ball, D. L. (1995). Transforming pedagogy: Classrooms as mathematical communities. A response to Timothy Lensmire and John Pryor. *Harvard Educational Review, 65*(4), 670–677.

Bandura, A. (1997). *Self-efficacy: The exercise of control.* New York: W.H. Freeman.

Banilower, E., & Shimkus, E. (2004). *Professional development observation study.* Chapel Hill, NC: Horizon Research.

Borko, H. (2004). Professional development and teacher learning: Mapping the terrain. *Educational Researcher, 33*(8), 3–15.

Cakmakci, G. (2012). Promoting pre-service teachers' ideas about nature of science through educational research apprenticeship. *Australian Journal of Teacher Education, 37*(2), 114–135.

Carey, N., & Frechtling, J. (1997). *Best practice in action: Follow-up survey on teacher enhancement programs.* Arlington, VA: National Science Foundation.

Carlsen, W. S. (1991). Questioning in classrooms: A sociolinguistic perspective. *Review of Educational Research, 61*(2), 157–178.

Carlsen, W. S. (1993). Teacher knowledge and discourse control: Quantitative evidence from novice biology teachers' classrooms. *Journal of Research in Science Teaching, 30*(5), 471–481.

Cazden, C. B. (2001). *Classroom discourse. The language of teaching and learning* (2nd ed.). Portsmouth, NH: Heinemann.

Chin, C. (2006). Classroom interaction in science: Teacher questioning and feedback to students' responses. *International Journal of Science Education, 28*(11), 1315–1346.

Cleland, J. V., Wetzel, K. A., Zambo, R., Buss, R. R., & Rillero, P. (1999). Science integrated with mathematics using language arts and technology: A model for collaborative professional development. *Journal of Computers in Mathematics and Science Teaching, 18*(2), 157–172.

Clough, M. P. (2006). Learners' responses to the demands of conceptual change: Considerations for effective nature of science instruction. *Science & Education, 15*(5), 463–494.

Cochran, K. F., DeRuiter, J. A., & King, R. A. (1993). Pedagogical content knowing: An integrative model for teacher preparation. *Journal of Teacher Education, 44*(4), 263–272.

Cochran-Smith, M., & Lytle, S. L. (1999). Relationships of knowledge and practice: Teacher learning in communities. *Review of Research in Education, 24*(1), 249–305.

Cohen, D. K., & Hill, H. C. (2001). *Learning policy: When state education reform works.* New Haven, CT: Yale University Press.

Cronin-Jones, L. L. (1991). Science teacher beliefs and their influence on curriculum implementation: Two case studies. *Journal of Research in Science Teaching, 28*(3), 235–250.

Czerniak, C. M. (1990). *A study of self-efficacy, anxiety, and science knowledge in pre-service elementary teachers.* A paper presented at the annual meeting of the National Association of Research in Science Teaching, Atlanta, GA.

Darling-Hammond, L. (1997). *Doing what matters most: Investing in quality teaching.* New York: National Commission on Teaching and America's Future.

Desimone, L. M. (2009). Improving impact studies of teachers' professional development: Toward better conceptualizations and measures. *Educational Researcher, 38*(3), 181–199.

Desimone, L. M., Garet, M., Birman, B., Porter, A., & Yoon, K. S. (2002). How do district management and implementation strategies relate to the quality of the professional development that districts provide to teachers? *Teachers College Record, 104*(7), 1265–1312.

Dorion, K. R. (2009). Science through drama: A multiple case exploration of the characteristics of drama activities used in secondary science lessons. *International Journal of Science Education, 31*(16), 2247–2270.

Driver, R., Leach, J., Millar, R., & Scott, P. (1996). *Young people's images of science.* Buckingham, UK: Open University Press.

Duschl, R. A. (2000). Making the nature of science explicit. In R. Millar, J. Leach, & J. Osborne (Eds.), *Improving science education: The contribution of research* (pp. 187–206). Philadelphia: Open University Press.

Edwards, D., & Westgate, D. P. G. (1994). *Investigating classroom talk.* London: Falmer Press.

Enochs, L. G., & Riggs, I. M. (1990). Further development of an elementary science teaching efficacy belief instrument: A preservice elementary scale. *School Science and Mathematics, 90*(8), 694–706.

Evens, M., Elen, J., & Depaepe, F. (2016). Pedagogical content knowledge in the context of foreign and second language teaching: A review of the research literature. *Porta Linguarum, 26,* 187–200.

Fernandez, C. (2014). Knowledge base for teaching and Pedagogical Content Knowledge (PCK): Some useful models and implications for teachers training. *Problems of Education in the 21st Century, 60,* 79–100.

Fernández-Balboa, J. M., & Stiehl, J. (1995). The generic nature of pedagogical content knowledge among college professors. *Teaching and Teacher Education, 11*(3), 293–306.

Ford, M. E. (1992). *Motivating humans: Goals, emotions, and personal agency beliefs.* Newbury Park, CA: Sage Publications.

Gaible, E. & Burns, M. (2005). Section 3: Models and best practices in teacher professional development. In G. Edmond & M. Burns, *Using technology to train teachers: Appropriate uses of ICT for teacher professional development in developing countries* (pp. 15–24). Washington, DC: infoDev / World Bank.

Garet, M. S., Porter, A. C., Desimone, L., Birman, B. F., & Yoon, K. S. (2001). What makes professional development effective? Results from a national sample of teachers. *American Educational Research Journal, 38*(4), 915–945.

Gerard, L. F., Varma, K., Corliss, S. B., & Linn, M. C. (2011). Professional development for technology-enhanced inquiry science. *Review of Educational Research, 81*(3), 408–448.

Gibson, S., & Dembo, M. H. (1984). Teacher efficacy: A construct validation. *Journal of Educational Psychology, 76*(4), 569–582.

Gross, D., Truesdale, C., & Bielec, S. (2001). Backs to the wall: Supporting teacher professional development with technology. *Educational Research and Evaluation, 7*(2–3), 161–183.

Grossman, P. L. (1990). *The making of a teacher: Teacher knowledge and teacher education.* New York: Teachers College Press.

Guskey, T. R. (1988). Teacher efficacy, self-concept, and attitudes toward the implementation of instructional innovation. *Teaching and Teacher Education, 4*(1), 63–69.

Guskey, T. R. (2000). *Evaluating professional development.* Thousand Oaks, CA: Corwin Press.

Hammer, D. (1995). Student inquiry in a physics class discussion. *Cognition and Instruction, 13*(3), 401–430.

Hammer, D. (2004). The variability of student reasoning, lectures 1–3. In E. Redish & M. Vicentini (Eds.), *Proceedings of the Enrico Fermi Summer School, Course CLVI* (pp. 279–340). Bologna: Italian Physical Society.

Hashweh, M. Z. (2005). Teacher pedagogical constructions: A reconfiguration of pedagogical content knowledge. *Teachers and Teaching, 11*(3), 273–292.

Hawley, W. D., & Valli, L. (1999). The essentials of effective professional development. In L. Darling-Hammond & G. Sykes (Eds.), *Teaching as the learning profession: Handbook of policy and practice* (pp. 127–150). San Francisco: Jossey-Bass.

Hill, H. C., Rowan, B., & Ball, D. L. (2005). Effects of teachers' mathematical knowledge for teaching on student achievement. *American Educational Research Journal, 42*(2), 371–406.

Hogan, K. (1999). Thinking aloud together: A test of an intervention to foster students' collaborative scientific reasoning. *Journal of Research in Science Teaching, 36*(10), 1085–1109.

Hollon, R. E., Roth, K. J., & Anderson, C. W. (1991). Science teachers' conceptions of teaching and learning. In J. Brophy (Ed.), *Advances in research on teaching* (Vol. 2, pp. 145–186). Greenwich, CT: JAI.

Kelly, G. J., & Crawford, T. (1997). An ethnographic investigation of the discourse processes of school science. *Science Education, 81*(5), 533–559.

Ketelhut, D. J., & Schifter, C. C. (2011). Teachers and game-based learning: Improving understanding of how to increase efficacy of adoption. *Computers & Education, 56*(2), 539–546.

Kind, V. (2009). Pedagogical content knowledge in science education: Perspectives and potential for progress. *Studies in Science Education, 45*(2), 169–204.

Koballa, T. R., Gräber, W., Coleman, D., & Kemp, A. C. (1999). Prospective teachers' conceptions of the knowledge base for teaching chemistry at the gymnasium. *Journal of Science Teacher Education, 10*(4), 269–286.

Lawless, K. A., & Pellegrino, J. W. (2007). Professional development in integrating technology into teaching and learning: Knowns, unknowns, and ways to pursue better questions and answers. *Review of Educational Research, 77*(4), 575–614.

Leach, J., Hind, A., & Ryder, J. (2003). Designing and evaluating short teaching strategy about the epistemology of science in high school classroom. *Science Education, 87*(6), 831–848.

Lederman, N. G. (1999). Teachers' understanding of the nature of science and classroom practice: Factors that facilitate or impede the relationship. *Journal of Research in Science Teaching, 36*(8), 916–929.

Lederman, N. G. (2007). Nature of science: Past, present and future. In S. A. Abell & N. G. Lederman (Eds.), *Handbook of research on science education* (pp. 831–879). London: Lawrence Erlbaum Associates.

Lemke, J. L. (1990). *Talking science: Language, learning & values*. Norwoord, NJ: Ablex.

Linn, M. C., Davis, E. A., & Bell, P. (Eds.). (2004). *Internet environments for science education* (pp. 29–46). Mahwah, NJ: Lawrence Erlbaum Associates.

Louca, L., & Zacharia, Z. (2007). Nascent abilities for scientific inquiry in elementary science. In S. Vosniadou, D. Kayser, & A. Protopapas (Eds.), *The proceedings of EuroCogSci07.The European Cognitive Science Conference 2007* (pp. 53–58). East Sussex, UK: Lawerence Erlbaum Associates.

Louca, T. L., Tzialli, D., & Zacharia, Z. (2012). Identification, interpretation–evaluation, response: A framework for analyzing classroom-based teacher discourse in science. *International Journal of Science Education, 34*(12), 1823–1856.

Loucks-Horsley, S., Hewson, P. W., Love, N., & Stiles, K. (1998). *Designing professional development for teachers of science and mathematics*. Thousand Oaks, CA: Corwin Press.

Loucks-Horsley, S., & Matsumoto, C. (1999). Research on professional development for teachers of mathematics and science: The state of the scene. *School Science and Mathematics, 99*(5), 258–271.

Loughran, J., & Gunstone, R. (1997). Professional development in residence: Developing reflection on science teaching and learning. *Journal of Education for Teaching, 23*(2), 159–179.

Loughran, J. J., Berry, A., & Mulhall, P. (2006). *Understanding and developing science teachers' pedagogical content knowledge*. Rotterdam, The Netherlands: Sense Publishers.

Magnusson, S., Krajcik, J., & Borko, H. (1999). Nature, sources and development of pedagogical content knowledge for science teaching. In J. Gess-Newsome & N. G. Lederman (Eds.), *Examining pedagogical content knowledge* (pp. 95–132). Dordrecht, The Netherlands: Kluwer.

Marks, R. (1990). Pedagogical content knowledge: From a mathematical case to a modified conception. *Journal of Teacher Education, 41*(3), 3–11.

Martin, J. R. (1993). Literacy in Science: Learning to handle text as technology. In M. A. K. Halliday & J. Martin (Eds.), *Writing Science: Literacy and discursive power* (pp. 166–202). Pittsburgh, PA: University of Pitts-burg Press.

May, D. B., Hammer, D., & Roy, P. (2006). Children's analogical reasoning in a 3rd-grade science discussion. *Science Education, 90*(2), 316–330.

McLaughlin, M., & Talbert, J. (1993). *Contexts that matter for teaching and learning.* Stanford, CA: Stanford University.

McSharry, G., & Jones, S. (2000). Role-play in science teaching and learning. *School Science Review, 82*(298), 73–82.

Minner, D. D., Levy, A. J., & Century, J. (2010). Inquiry-based science instruction – What is it and does it matter? Results from research synthesis from years 1984 to 2002. *Journal of Re-search in Science Teaching, 47*(4), 474–496.

Minstrell, J. A., & Van Zee, E. H. (Eds.). (2000). Inquiring into inquiry: Learning and teaching in science. Washington, DC: AAAS.

Mortimer, E. F. (1995). Addressing obstacles in the classroom: An example from the theory of matter. Paper presented at the European Conference on Research in Science Education. Leeds.

Mortimer, E. F., & Scott, P. (2003). *Meaning making in science classrooms.* Milton Keynes: Open University Press.

MTG. (2007). *Mind the Gap: Learning, Teaching, Research and Policy in Inquiry-Based Science Education*, Grant Agreement Number 217725, European Commission.

National Research Council [NRC]. (1996). *National science education standards.* Washington, DC: National Academy Press.

National Research Council [NRC]. (2007). *Taking science to school: Leaning and teaching science in grades K-8.* Washington, DC: The National Academies Press.

National Science Teachers Association [NSTA]. (2000). NSTA position statement on the nature of science. Retrieved January 12, 2011, from http://www.nsta.org/about/positions/natureof-science.aspx

Nilsson, P., & Vikström, A. (2015). Making PCK explicit-capturing science teachers' pedagogical content knowledge (PCK) in the science classroom. *International Journal of Science Education, 37*(17), 2836–2857.

O'Toole, J., & Dunn, J. (2002). *Pretending to learn: Helping children learn through drama.* Longman, an imprint of Pearson Education Australia.

OECD. (2009). *Teaching and learning international survey creating effective teaching and learning environments: First results from TALIS.* Paris: OECD.

Opfer, V. D., & Pedder, D. (2011). Conceptualizing teacher professional learning. *Review of Educational Research, 81*(3), 379–407.

Osborne, J., & Dillon, J. (2008). *Science education in Europe: Critical reflections* (Vol. 13). London: The Nuffield Foundation.

Osborne, J., Erduran, S., & Simon, S. (2004). Enhancing the quality of argumentation in school science. *Journal of Research in Science Teaching, 41*(10), 994–1020.

Penuel, W. R., Fishman, B., Gallagher, L., Korbak, C., & Lopez-Prado, B. (2008). Is alignment enough? Investigating the effects of state policies and professional development on science curriculum implementation. *Science Education, 93*(4), 656–677.

Penuel, W. R., Fishman, B., Yamaguchi, R., & Gallagher, L. P. (2007). What makes professional development effective? Strategies that foster curriculum implementation. *American Educational Research Journal, 44*(4), 921–958.

Penuel, W. R., & Yarnall, L. (2005). Designing handheld software to support classroom assessment: Analysis of conditions for teacher adoption. *Journal of Technology, Learning and Assessment, 3*(5), 1–46.

Phillips, J. (2003). Powerful learning: Creating learning communities in urban school reform. *Journal of Curriculum and Supervision, 18*(3), 240–258.

Remillard, J. T., & Geist, P. (2002). Supporting teachers' professional learning though navigating openings in the curriculum. *Journal of Mathematics Teacher Education, 5*(1), 7–34.

Russ, R., Scherr, E. R., Hammer, D., & Mikeska, J. (2008). Recognizing mechanistic reasoning in scientific inquiry. *Science Education, 92*(3), 499–525.

Sadeh, I., & Zion, M. (2009). The development of dynamic inquiry performances within an open inquiry setting: A comparison to guided inquiry setting. *Journal of Research in Science Teaching, 40*(10), 1137–1116.

Scott, P., & Ametller, J. (2007). Teaching science in a meaningful way: Striking a balance be-tween 'opening up' and 'closing down' classroom talk. *School Science Review, 88*(324), 77–83.

Shulman, L. S. (1986). Those who understand: Knowledge growth in teaching. *Educational Researcher, 15*(2), 4–14.

Shulman, L. S. (1987). Knowledge and teaching: Foundations of the new reform. *Harvard Educational Review, 57*(1), 1–23.

Smith, D. (1999). Changing our teaching: The role of pedagogical content knowledge in elemen-tary science. In J. Gess-Newsome & N. Lederman (Eds.), *Examining pedagogical content knowledge: The construct and its implications for science teacher education* (pp. 163–198). Dordrecht, The Netherlands: Kluwer Academic Publishers.

Solomon, J. (1994). The rise and fall of constructivism. *Studies in Science Education, 23*, 1–19.

Sprod, T. (1998). I can change your opinion on that: Social constructivist whole class discussions and their effect on scientific reasoning. *Research in Science Education, 28*(4), 463–480.

Strahan, D. (2003). Promoting a collaborative professional culture in three elementary schools that have beaten the odds. *The Elementary School Journal, 104*(2), 127–146.

Supovitz, J. A., & Zeif, S. G. (2000). Why they stay away. *Journal of Staff Development, 21*(4), 24–28.

Tamir, P. (1988). Subject matter and related pedagogical knowledge in teacher education. *Teaching and Teacher Education, 4*(2), 99–110.

Thomas, G., Wineburg, S., Grossman, P., Myhre, O., & Woolworth, S. (1998). In the company of colleagues: An interim report on the development of a community of teacher learners. *Teaching and Teacher Education, 14*(1), 21–32.

Tobin, K., Tippins, D. J., & Gallard, A. J. (1994). Research on instructional strategies for teaching science. In D. Gabel (Ed.), *Handbook of research on science teaching and learning* (pp. 45–93). New York, NY: Macmillan Publishing Company.

Van Driel, J. H., Verloop, N., & de Vos, W. (1998). Developing science teachers' pedagogical con-tent knowledge. *Journal of Research in Science Teaching, 35*(6), 673–695.

van Zee, E. H. (2000). Analysis of a student-generated inquiry discussion. *International Journal of Science Education, 22*, 115–142.

van Zee, E. H., & Minstrell, J. (1997). Using questioning to guide student thinking. *The Journal of the Learning Sciences, 6*(2), 227–269.

Veal, W. R., & MaKinster, J. G. (1999). Pedagogical content knowledge taxonomies. *Electronic Journal of Science Education, 3*(4) From http://unr.edu/homepage/crowther/ejse/ejsev3n4.html

Wayne, A. J., Yoon, K. S., Zhu, P., Cronen, S., & Garet, M. S. (2008). Experimenting with teacher professional development: Motives and methods. *Educational Researcher, 37*(8), 469–479.

Wells, G. (1999). *Dialogic inquiry. Towards a sociocultural practice and theory of education.* Cambridge, UK: Cambridge University Press.

Williams, M. (2008). Moving technology to the center of instruction: How one experienced teacher incorporates a Web-based environment over time. *Journal of Science Education and Technology, 17*(4), 316–333.

Woolfolk, A. E., & Hoy, W. K. (1990). Prospective teachers' sense of efficacy beliefs about con-trol. *Journal of Educational Psychology, 82*, 81–91.

Yamagata-Lynch, L. C. (2003). How a technology professional development program fit into the work lives of teachers. *Teaching and Teacher Education, 19*(6), 591–607.

Index

A
Abell, S., 105
Aksela, M., 188
Albert, T.K., 175
Alexander, R.J., 123
Ametller, J., 123, 269
Anderman, L.H., 89
Andersen, H.M., 87
Anderson, R., 6, 18, 187
Anderson, T., 4
Arnold, J.C., 4
Ashkenazi, G., 73
Ausubel, D.P., 2
Autonomy, 9, 28, 30, 35, 36, 38, 39, 78, 90,
    91, 96–100, 107, 187, 188, 191, 197,
    199, 216, 245–247, 250–256,
    264, 265, 267
Awareness, 5, 6, 9–12, 47, 88, 95–96,
    101, 132, 232, 233, 256, 267

B
Badreddine, Z., 247
Bakhtin, M., 227, 236
Banchi, H., 166
Bandura, A., 43, 107, 109, 110, 117, 268
Barnes, D., 226
Barney, E.C., 49
Barron, L., 175
Batagelj, V., 111
Bell, B., 143
Bell, P., 141, 224
Bell, R., 166
Biology Olympiad (BiO), 206, 211, 213

Black, P., 81
Bransford, J.D., 167
Breyfogle, B.E., 215, 216
Brito, A., 158
Brousseau, G., 246
Brown, A.L., 13
Bruner, J., 1
Bruun, J., 111
Bungum, B., 31, 34, 36, 264
Bybee, R.W., 28

C
Cakmakci, G., 137–159
Campione, J.C., 13
Carlsen, W., 36
Century, J., 186
Chinn, C.A., 11, 13
Clark, R.E., 13
Cochran, K.F., 272
Cogan, J.G., 10
Colburn, A., 15
Collaborative learning, 1, 165, 166,
    168–171, 175–180
Communicative approach, 121, 123–129,
    131, 225–227, 229–233, 235, 266,
    268–269, 275
Concept networks, 106, 109–113, 116–118
Constantinou, C.P., 1
Cowie, B., 143
Cross, D., 247
Cross-curricular content, 164–167, 169–171,
    177–180
Cross-curricular teaching, 31, 172

© Springer International Publishing AG, part of Springer Nature 2018
O. E. Tsivitanidou et al. (eds.), *Professional Development for Inquiry-Based
Science Teaching and Learning*, Contributions from Science Education
Research 5, https://doi.org/10.1007/978-3-319-91406-0

D
Dapkus, D., 163–180
Davis, E.A., 141, 224
Dawes, L., 227
De Simone, C., 175
Deci, E.L., 95
DeRuiter, J.A., 272
Design and technology, 31–33, 36–38, 264, 266
Desimone, L., 273
Dewey, J., 1, 2, 43, 166
Dialogic inquiry, 123–126, 130, 223–237, 269
Dialogic teaching, 122, 130, 226, 266, 268–269, 275
Didactic tool, 213
Dillon, J., 87
Dolin, J., 111, 267, 268
Drama, 65–81, 133, 265
Duit, R., 44, 45
Duncan, R.G., 13
Dunn, J., 67, 75

E
Edelson, D.C., 224
Educational innovation, 12, 18, 19, 189
Educational reconstruction, 44, 45
Edwards, S., 175
Eliot, C., 43
Embodied learning, 77
European educational policies, 7, 15
Evans, R., 111, 267, 268
Explicit-reflective Nature of Science (NOS) instruction, 142, 144, 151

F
Foldevi, M., 165, 168
Ford, M.E., 89, 268
Formative assessment, 7, 13, 15, 66, 80–81, 138, 142–145, 148–152, 231

G
Gago, J.M., 17
Gallagher, L.P., 188
Gibson, H.L., 10
Gillies, R.M., 179
Gok, T., 163, 164, 167
Gordin, D.N., 224
Graf, J., 198
Grangeat, M., 7, 18

Gropengießer, H., 44
Grossman, P.L., 272
Gyllenpalm, J., 224

H
Hammer, M., 175
Hamo, A., 60
Hansson, A., 165
Hargreaves, A., 31
Hattie, J., 80
Haug, B.S., 81
Hayes, J.R., 167
Holmberg, S-O., 224

I
Inquiry, 1–19, 27–39, 46, 48, 66–75, 77, 78, 80, 81, 88, 90, 92, 96, 100, 105, 121–132, 138–141, 146, 150, 152, 164, 166, 170, 177–179, 205–219, 223–237, 261–275
Inquiry-based learning (IBL), 8–12, 14, 15, 17, 28, 39, 66, 69, 73, 74, 77, 81, 124, 126, 130, 166, 263–265, 268
Inquiry based science teaching (IBST), 4, 6–9, 18, 116, 185, 197, 206, 223, 224, 226, 245, 261–275
Inquiry based science teaching and learning (IBST/L), 4–6, 8–15, 43, 44, 60, 81, 106–110, 116–118, 125, 137–141, 150, 185–200, 224, 229, 237, 261–275
Inquiry on biology, 206
Interactional graphic, 121, 127–132

J
Jonassen, D.H., 167
Jones, S., 67, 68
Jørgensen, E.C., 31, 34
Juuti, K., 185, 188

K
Katedralskole, A., 88
Kattman, U., 44
Kellert, S.R., 49
Kenedy, D., 175
Kilpatrick, W., 43
King, R.A., 272
Kirschner, P.A., 13
Klaassen, K., 44
Klus-Stańska, D., 43

Koh, K., 175
Kolčavová, Z., 210
Komorek, M., 44
Kremer, K., 4
Krogh, L.B., 88, 89

L

Laursen, S., 198
Lavonen, J., 186, 188
Leake, V.S., 89
Learning, 1, 27, 43, 66, 90, 106,
    122, 138, 163, 185, 205,
    223, 245, 262
Lederman, N., 105
Lederman, N.G., 139
Lehesvuori, S., 128
Levy, A.J., 186
Lijnse, P., 44
Linn, M.C., 141, 224
Liston, C., 198
Looney, J.W., 15
Loucks-Horsley, S., 113
Loukomies, A., 185

M

Macfarlane-Dick, D., 143
Magnusson, S., 272
Malhotra, B.A., 11
Malone, J., 29
Marks, R., 272
Mattsson, B., 165
Mayer, J., 4
McComas, W.F., 139
McPhee, A., 175
McSharry, G., 67, 68
Meisalo, V., 188
Mercer, N., 227
Metallidou, P., 167
Millar, R., 60
Minner, D.D., 4, 186, 187, 190, 191,
    193–195, 197, 198
Mintzes, J.J., 49, 59
Moate, J., 121
Mohamed, M.E.S., 175
Mork, S.M., 81
Mortimer, E.F., 123, 126, 225, 226, 269
Motivation, 7, 12, 15, 17, 31, 35, 44, 65,
    87–101, 118, 123, 141, 143, 150,
    185, 187, 189–191, 194–197, 199,
    206, 208–212, 217, 264–266,
    268, 274, 275
Mrvar, A., 111

N

Nature of science (NOS), 9, 10, 12, 18, 27, 28,
    137–159, 187, 208, 245, 253,
    266, 269, 275
Niaz, M., 154, 158, 159
Nicol, D.J., 143
Nooy, W., 111

O

O'Toole, J., 67, 75
Ødegaard, M., 81
Olson, S., 113
Orlando, 226
Osborne, J., 60, 87, 236
Østern, A.L., 73

P

Papáček, M., 205
Pea, R.D., 224
Pečiuliauskienė, P., 163
Pedagogical content knowledge (PCK),
    188, 217, 262, 270–273, 275
Peleg, R., 80
Penuel, W.R., 198
Petr, J., 214
Piaget, J., 2
Popular media, 141
Pre-service science teachers, 165, 176, 180
Preston-Sabin, J., 175
Problem-based learning (PBL), 7, 16, 44, 107,
    163–168, 170–173, 206, 208
Problem-solving abilities, 165, 166, 171,
    173–176, 178, 180
Professional development, 8, 11, 13, 15, 18,
    30, 33, 78, 79, 81, 87, 88, 92, 93, 100,
    101, 107, 108, 116, 118, 131, 132, 150,
    185–191, 198, 199, 211, 212, 217,
    223–237, 255, 261, 263, 267, 268,
    270–275
Prokop, M., 45
Prokop, P., 45

R

Rahman, S., 175
Ratinen, I., 128
Reflection, 1, 30, 36, 70, 73, 79, 88, 92–94,
    96, 97, 99, 116, 129, 131, 143, 175,
    185, 189, 191, 226, 227, 229–237, 264
Reflective writing, 88, 94, 101, 267
Rennie, L., 29
Rocard, M., 3, 16

Rodgers, C., 189
Rodrguez, A., 158
Ryan, R.M., 95
Rybska, E., 43–61

S
Savery, J.R., 164
Sawyer, K., 188, 189, 198, 199
Scerri, E., 154, 158
School-based development, 27, 37–39
Schroeder, C.M., 10
Schwab, J.J., 2
Science competitions, 208
Science education, 1–16, 18–19, 27, 29, 43,
    65–68, 71, 73–77, 79, 81, 122, 123,
    125, 126, 137, 138, 166, 168, 169,
    172, 178, 179, 186, 205–209, 237,
    262, 263, 265, 266, 268, 270, 275
Science Olympiad (SciO), 205–208, 216
Science teaching, 66
Scientific literacy, 6, 16, 59, 65, 105–118, 137,
    139, 141, 227, 234, 265–268, 270, 275
Scott, P., 123, 126, 225, 226, 236, 269
Selander, S., 67
Self-efficacy, 89–91, 97, 100, 107,
    266–268, 275
Shepard, L.A., 81
Sheridan, S., 208
Sherin, M.G., 96
Shulman, L.S., 13, 88, 271
Singer, S.R., 87
Skerrett, A., 31
Sørvik, G.O., 81
Stazinski, W., 208
Stein, B.S., 167
Stepwise problem solving strategies, 172
Stipek, D., 95
Strømme, A., 73
Stuchlíková, I., 206
Student motivation, 65, 87, 90, 94,
    187, 191, 265
Svendsen, B., 81
Sweller, J., 13

T
Tamir, P., 60, 272
Teacher education, 18, 71, 72, 79, 132, 143,
    144, 150, 206, 216–217, 236, 271
Teacher professional development (TPD), 15,
    78, 92, 93, 107, 108, 116, 118, 132,
    150, 188, 205, 207, 217, 223–237,
    262, 268, 270, 272–275

Teachers as professionals, 30–31, 188, 189
Teaching, 1, 27, 43, 66, 87, 106, 121,
    137, 153, 164, 185, 205, 223,
    245, 262
Teaching-learning sequence (TLS),
    45–47, 126, 264
Theory-to-practice, 87
Thiery, H., 198
Thompson, T.L., 59
Tiberghien, A., 247
Timperley, H., 80
Tobin, W., 30
Todds, F., 226
Tsabari, A.B., 66
Tsivitanidou, O.E., 1
Tteacher education, 245–257
Tubin, D., 30, 37
Tunnicliffe, S.D., 45
Tyack, D., 30

V
Venville, G., 29, 36
Video club, 88, 93, 96–101
Viiri, J., 128
Vygotsk, L., 2
Vygotsky, L., 69, 70, 227, 236

W
Wallace, J., 29
Wegerif, R., 227
Wellington, J., 236
Wells, G., 224, 227
Wickström, P-O., 224
Williams, P., 208
Worrall, J., 158
Writing in dialog, 229, 235, 236

Y
Yalaki, Y., 137–159
Yamaguchi, R., 188
Yamat, H., 175
Yassin, S.F.M., 175
Yeager, C., 187, 190, 198
Yen, C.F., 49
Yeung, K.H., 175

Z
Ziman, J., 29
Zoller, U., 10
Zsoldos–Marchis, I., 179

Printed in the United States
Bookmasters